THE FUTURE OF CHANGE

THE FUTURE OF CHANGE

How Technology Shapes Social Revolutions

Ray Brescia

CORNELL UNIVERSITY PRESS ITHACA AND LONDON

First published 2020 by Cornell University Press

Library of Congress Cataloging-in-Publication Data

Names: Brescia, Ray, author.
Title: The future of change : how technology shapes social revolutions / Ray Brescia.
Description: Ithaca : Cornell University Press, 2020. | Includes bibliographical references and index.
Identifiers: LCCN 2019029567 (print) | LCCN 2019029568 (ebook) | ISBN 9781501748110 (hardcover) | ISBN 9781501748127 (epub) | ISBN 9781501748134 (pdf)
Subjects: LCSH: Technological innovations—Social aspects—United States—History. | Social change—United States—History. | Social movements—United States—History. | Equality—United States—History. | United States—Social conditions—1945-
Classification: LCC HM846 .B745 2020 (print) | LCC HM846 (ebook) | DDC 303.40973—dc23
LC record available at https://lccn.loc.gov/2019029567
LC ebook record available at https://lccn.loc.gov/2019029568

To Amy and Leo
My Love and My Life

Thus the most democratic country on earth is found to be, above all, the one where men in our day have most perfected the art of pursuing the object of their common desires in common and have applied this new science to the most objects. Does this result from an accident or could it be that there in fact exists a necessary relation between associations and equality?

Alexis de Tocqueville

Contents

Preface

In many ways, the ideas in this book began to emerge in my early days as a lawyer for tenant organizing groups in Harlem and Washington Heights in New York City, while I was working at the Legal Aid Society of New York in the early 1990s. During my years there, the leaders of the tenant associations I represented—Daryl Edwards, Byron Utley, Reina Lendor, Gyreain Privette, Frankie Clark, and countless others—taught me almost everything I know about community organizing and social change. Later, community leaders like Wing Lam and Mary Dailey helped educate me even further. I have had the distinct pleasure and honor to work with a large number of community leaders over the years, but there are simply too many to name here. I also have the good fortune to call many colleagues my friends, mentors, and allies. Although I regret I cannot name them all here either, I want to mention Molly Biklen, Harvey Epstein, April Herms, Carmen Huertas-Noble, Doug Lasdon, Megan Lewis, Tony Lu, Andrew Kashyap, Gowri Krishna, Annie Lai, Serge Martinez, Laine Romero-Alston, Anika Singh Lemar, Saba Waheed, David Weinraub, John Whitlow, John Wright, and Haeyoung Yoon. This book is a reflection of and testament to much of our work together.

The support I received from Albany Law School was central to this project. The law school's president and dean, Alicia Ouellette, has been supportive both professionally and personally. In addition, treasured colleagues offered helpful guidance by reading drafts of excerpts, including Andrew Ayers, Ted DeBarbieri, Steve Gottlieb, Keith Hirokawa, and Sarah Rogerson. My former colleague and law school classmate Tim Lytton offered an incredible degree of support and guidance, from the earliest days of the writing process and even after leaving the school to join the faculty at Georgia State University School of Law. His friendship and advice were critical to the book's creation. In addition, research assistants Lindsey Dodd, Heather Hage, Mary Ann Krisa, and Lauren McCluskey provided significant help in countless ways, for which I am grateful. My legal assistant at Albany Law, Sherri Meyer, was tireless and always of good cheer, no matter how much work I threw her way. The leadership and staff of the Albany Law School library, especially David Walker and Pegeen Lorusso and students Alex-Marie Baez, Claire Burke, and Victoria Soracco, were all tremendously helpful throughout this process, providing support and assistance whenever it was needed. Similarly, the school's Department of Information Technology Services met my technological needs with good cheer and tremendous patience. The staff at the New York State Library also offered critical assistance in locating some of the nineteenth-century materials scattered throughout the work.

Others outside of Albany Law School were also deeply helpful, including the members of the Clinical Theory Workshop organized at New York Law School, especially Jason Parkin, David Reiss, Ian Weinstein, and the late Stephen Ellmann, whose early scholarship on representing grassroots organizations in a mindful and client-focused way continues to inspire generations of lawyers and scholars.

Countless friends offered a great deal of guidance and support, including Mike Wishnie, Graham Boyd, Harold Hongju Koh, Paul Sonn, Susan Plum, Rosemary Queenan, Gates Garrity-Rokous, Beth Garrity-Rokous, Brandt Goldstein, Mark Napier, Nicole Theodosiou, Cynthia Fox, Bob Solomon, Joyce Gelb, Richard Pinner, Adam Bramwell, Charles Chesnut, Chris Coons, Kurt Peterson, and James Esseks. One of my friends and mentors, Michael Ratner, who did not live to see this book in print, contributed to my development as a lawyer for grassroots groups as much as anyone. He continues to be an inspiration to many, myself included.

The journalist and author Jonathan Rosen provided early editorial guidance. I am also grateful to Kate Babbitt for her editorial assistance and insightful research ideas that made the book better and more focused.

I also thank everyone I interviewed for the book: Ady Barkan, Emily Comer, Rosie Hidalgo, Alan Jenkins, Loraina Lopez Masoumi, Amy Mello, Jay O'Neal, Jeanine Pearce, Pat Reuss, Sharon Stapel, Leigh Shelton, Richard Viguerie, and Evan Wolfson. I am grateful to several others who provided insights, though their interviews did not make it into the book (because of space limits), including Ben Berkowitz, Cory Booker, Brad Lander, Alexa Kasden, Andrew Kennedy, and Kam Lasater. Their thoughts and ideas certainly animate this work as much as those whose words appear in it.

No one is more responsible for this work coming to fruition than Emily Andrew from Cornell University Press. She identified many problems in earlier drafts, and yet she believed I just might be on to something. Her patience, guidance, support, and constructive criticism were invaluable. She is everything an author could want in an editor, and I am grateful to have had her in my corner. The comments and criticisms of anonymous reviewers strengthened the book immeasurably. I have done my best to respond to all of their critiques.

My immediate and extended family provided great moral—and editorial—support. My brother-in-law, Richard Marsico, read multiple drafts and provided lots of useful and constructive criticism to help improve the manuscript.

My son, Leo, continues to be an inspiration. My hope is that the world this book envisions is one in which he will live.

My spouse, Amy Barasch, put up with me for the eight long years it has taken to bring these ideas and this book into the world. We celebrated our twentieth wedding anniversary in 2019 and have been together since 1996. No one knows more about my thoughts on social change and community organizing than she does. It was our work serving low-income communities of color that brought us together in the first place. She's a great lawyer, advocate, leader, and writer, and her thoughtful reviews of earlier drafts made them stronger. More important, she has made me feel truly loved throughout, and nothing is possible without love.

While many have contributed to this book, any omissions or flaws are solely my responsibility. I hope those who read it will do so with generosity and receive it in the spirit with which it was written: as a humble effort to present a hopeful vision for a more democratic, more participatory, more progressive, more human, and more humane future.

Abbreviations

ABC	American Broadcasting Company
ACLU	American Civil Liberties Union
AEI	American Enterprise Institute
AFL-CIO	American Federation of Labor and Congress of Industrial Organizations
AIDS	acquired immunodeficiency syndrome
AVP	Anti-Violence Project
AWSA	American Woman Suffrage Association
CPD	Center for Popular Democracy
ERA	Equal Rights Amendment
HIV	human immunodeficiency virus
LA Gay & Lesbian Center	Los Angeles Gay & Lesbian Center
LAANE	Los Angeles Alliance for a New Economy
LDF	Legal Defense and Educational Fund, the legal arm of the NAACP
LGBTQ	lesbian, gay, bisexual, transgender, and questioning
MSNBC	news channel whose name combines "Microsoft" and the acronym for "National Broadcasting Company"
NAACP	National Association for the Advancement of Colored People
NFWA	National Farm Worker Association
NOW	National Organization for Women
NWSA	National Woman Suffrage Association
PD	prisoner's dilemma
PEIA	Public Employees Insurance Agency
PRLDF	Puerto Rican Legal Defense & Education Fund
RNC	Republican National Committee
UNITE HERE	Organization formed through the merger of the Union of Needletrades, Industrial and Textile Employees and the Hotel Employees and Restaurant Employees International Union
VAWA	Violence Against Women Act
WWII	World War II
YAF	Young Americans for Freedom

THE FUTURE OF CHANGE

INTRODUCTION

Making Social Change Happen

The American Legion faced its toughest battle yet. For several years, as the Second World War continued to rage in the European and Pacific theaters, the Legion had worked to convince the U.S. Congress to pass what would come to be known as the G.I. Bill, a package of benefits for service men and women returning from the war. To accomplish this, it activated members from its network of over ten thousand local chapters spread throughout the country. It used that network to mobilize public sentiment to support the initiative, encourage the Roosevelt administration to back it, and pressure Congress to approve it. Returning veterans stood to benefit from the groundbreaking program because it would open the doors to college and vocational training and offer many the opportunity to own homes. The Legion harnessed modern communications technologies to promote its message, using radio addresses and short film clips shown in movie theaters. It embarked on letter-writing campaigns and engaged in advocacy with newspaper editorial boards in its efforts to garner public support for the bill. It communicated with its members spread throughout the nation using all manner of communication: the mail, the telegraph, and the telephone. But despite all of these modern means of communication that enabled the Legion to build and coordinate a far-flung network of supporters, its efforts might have been for naught, all because of a hunting trip that occurred mostly beyond the reach of these technologies.

Even though the legislation enjoyed unanimous support in Congress, it still faced the very real possibility of defeat. Despite its efforts managing tens of thousands of volunteers in a network that spanned the nation, the outcome of the legislation in its then-current and generous form hung in the balance because of one person. The Legion had lost track of a single congressman: a swing vote on a crucial committee that controlled the fate of the legislation.

Years of painstaking work came to a head in June 1944, just days after the invasion of Normandy by Allied troops. Members of a select congressional committee in charge of the legislation gathered for an up-or-down vote on the benefits package. After different versions of the bill passed unanimously in the House of Representatives and Senate, Congress convened a joint conference of senators and representatives to reconcile the differences between the two chambers' versions of the bill. In order to pass a single version of the legislation, a majority of each of the chambers' representatives on the committee had to approve it. The members of the committee from the House were deadlocked over a few critical issues. One congressman, a Democrat from Georgia named John Gibson, could decide the outcome. Without his vote, the legislation would stall in Congress. Gibson supported the legislation, but he was not in Washington on the eve of the critical vote. Making matters worse, he could not be found. He was supposed to be at home recuperating from an illness. Even though he was needed in Washington the next day, he had apparently felt good enough to go on a hunting trip in his home state and was unreachable.

The Legion harnessed all means available to track Gibson down. Telephone calls to his home revealed his absence. By one account, when the telephone operator learned of the Legion's efforts, she disclosed that her husband was fighting in northern France and promised to call Gibson's home every five minutes until she found him. The Legion enlisted the support of local radio stations in Georgia, which asked their audiences for anyone with knowledge of the congressman's whereabouts to call Washington. Police officers stopped cars on highways in their search for the missing lawmaker. Reporters at the *Atlanta Journal-Constitution* called throughout the state to find him.

Through this statewide manhunt, the Legion finally located Gibson, and he was whisked back to the nation's capital by plane in the dead of night to cast his critical vote the next morning. Evoking Paul Revere's mythic exploits on the eve of the events at Lexington and Concord, Gibson's trip was dubbed the "Midnight Ride."[1] Once Gibson appeared in time to register his support for the bill, passage was assured.

The story of the passage of the G.I. Bill reveals how an adaptive grassroots network utilized all the media technologies available to it at the time in creative ways—from the mail and the telegraph to the radio and the cinema—to promote a positive, inclusive message and bring about social change. Innovation in communications technologies created an opportunity for the American Legion;

it had at its disposal a vast array of tools to not just communicate with but also coordinate the efforts of its vast network of local chapters to promote adoption of the program. This connection between communications technology and a social movement is not accidental. U.S. history reveals the deep relationship between social change and innovation in the means of communication. When new ways of communicating have emerged, new social movements seem to have risen up in their wake. The rapid spread of the printing press in the New World in the mid-eighteenth century helped spawn a revolution, just as it had a century before in England. The growth of the postal service after the creation of the new American republic helped facilitate the emergence of the Second Great Awakening and other social movements in the early nineteenth century. The steam printing press supercharged the capacities of the abolitionist movement in the 1830s. The telegraph spread the word of the birth of the women's movement. The introduction of the telephone and transcontinental railroad helped launch the suffragettes and the Progressive Era. The radio helped garner support for the New Deal. The television advanced the cause of civil rights in the 1960s. Today, mobile technologies have exposed police brutality in new ways that helped launch the Black Lives Matter Movement.

It might be tempting to say that this apparent connection between new communications technologies and new social movements suggests that technology causes social change to happen. Some today might argue that social media launched the Arab Spring, the Tea Party Movement, and other grassroots efforts that have emerged in recent years. But social change is far more complicated than that, as is its relationship to advances in technology. In this book I examine the link between, on the one hand, innovations in communications technology and methods and, on the other, social movements that appear to have emerged in their wake. I strive not just to understand the many ways in which communications and social movements are connected, but also to identify the components of the successes and failures of these same movements that seem to have a symbiotic relationship to the technology that fuels them.

Today, there is a growing dissatisfaction with the state of the world and the United States. This is evidenced in not just the election of Donald Trump to the presidency, swept in on a tide of resentment and a flurry of tweets, but also in the growing opposition that has arisen as a result of his victory in the election of 2016. These uprisings are being fueled by a desire to spark change in the status quo, regardless of where on the political spectrum a movement's leaders may fall. Whether it is seeking to promote gun control or expand gun rights, improve working conditions of low-wage workers, or reduce deaths from opioid addiction, many are looking for strategies for bringing about social change. Those who wish to create such change today—a time when the ability to communicate and connect with others has improved dramatically—can learn from successful social movements of the past, particularly those that succeeded in times that have experienced similar advances in the means of communication.

Throughout U.S. history, technological and other changes to the means of communication have helped fuel social change. But technology alone does not bring about such change. A review of social movement successes at times when the means of communications changed throughout U.S. history also reveals the following: these social movements share other common features. An understanding of such features might generate a formula for social change in these periods following technological change. In order to understand what makes a social movement successful in moments when the means of communications change, I explore whether there are common elements of successful social movements in such moments in the past and those movements that are emerging today. In this work, I call such eras—when innovations and changes in the means of communication have helped spark social change—"social innovation moments": when groups have harnessed these new means of communication to help bring new social arrangements into the world. In constitutional theory, the legal scholar Bruce Ackerman has identified what he describes as "constitutional moments": times when both leaders and popular movements brought about transformations of our understanding of the U.S. Constitution.[2] Is it possible that technological change creates a similar social innovation moment, when advances in the means of communication capture the imagination of activists and nonactivists alike and encourage the growth and spread of social movements? Changes in the ability to communicate seem to create an environment in which social movements can emerge, embrace the new technology, and use it to advance the change they wish to see in the world. If we are indeed in such a moment today—as citizens have new tools at their disposal to communicate and coordinate their actions at lightning speed—could the history of and experiences in previous social innovation moments, and the successes of social movements in such moments, help reveal the ways in which advocates can advance social change today and tomorrow? It is the goal of this book to explore this question.

In this exploration, I draw from historical and theoretical research, primary research on several contemporary social movements, and personal experience as a lawyer and former community organizer to identify the essential elements of social movement success in moments of technological change. A "social movement" is traditionally defined as an organized effort to bring about social change.[3] Similarly, a "social revolution" occurs when there is a significant shift in the social structure.[4] Social movements often create social revolutions. For my purposes here, I will consider a successful social movement primarily one that leads to some significant shift in the legal infrastructure that governs the social problem the movement was attempting to solve; in other words, I will focus on social movements that have led to social revolutions. For example, the colonial revolt against the British Crown led to a monumental shift in the legal infrastructure governing the colonies. In addition, I consider other changes as successes: e.g., winning the right to vote for women in the U.S. and making de jure segregation by race illegal. I also look at the small wins secured by low-income people operating at the grassroots to try to tease out larger lessons from their victories, challenges, and defeats.

While identifying a significant shift in the legal infrastructure is one way to measure social change, there are also more subtle ways in which social movements can achieve success, and a change in social norms—which often comes with a shift in the legal infrastructure but by no means has to—can be no less important; in fact, such change is often more important than legal change. Nevertheless, such shifts in social norms are hard to measure, and even polling data on shifts in social norms are often suspect, as poll respondents may be less willing to admit that they have not adopted new norms as they evolve. I will thus draw from the measurable achievements of a number of social movements that emerged in social innovation moments throughout U.S. history to attempt to identify common themes, strategies, and tactics of such movements. Measuring social change is hard, in the end, and I will attempt to identify what appear as clear examples of social movement victories to help identify the components of those victories as a way to unlock the secrets of making social change happen in social innovation moments.

Yet there are some things this book does not do. It is not an analysis of the ethics of technology, the rise of the surveillance state, or a diatribe about the risks of social media.[5] It is also not a love letter to technology. My goals are far less ambitious. What I hope to do in this book is provide a realistic assessment of contemporary communications technologies and the promise they may hold for those interested in bringing about social change. I borrow from both the past and the present, from theory and lived experience, to try to make sense of the social innovation moment we are in today and the moments to come, to understand the relationship between communications technology and social change so that those looking to advance change have a clear-eyed view of the capacities, and potential risks, of harnessing technology to bring about such change. I strive to draw from the past while looking forward to the future, taking lessons from the experiences of social movements that have taken shape at similar times when new communications technologies have emerged in U.S. history to inform approaches to organizing in a new, digital age. It is my hope that a review of the successes of past social movements that have deployed communications technology in effective ways will help uncover not just the components of those successes but also the reasons for them. The lessons will then help those looking to bring about social change with those communications tools currently available to them and those that will inevitably emerge in the future.

Another thing I will mostly avoid is an exploration of the successes of what some would label conservative social movements. The main goal of this work is to test and assess the components of successful *progressive* social change movements, from the abolition of slavery to the contemporary campaign to promote marriage equality. While not all social movements are progressive, the ones I will mostly explore in this book are those that sought to promote liberal or progressive causes simply because these are the types of causes that often seem to appear to emerge in the wake of advances in the means of communication. Moreover, I will also admit that they are the ones that are of greater interest to me. While the successes of social movements generally might help unlock the secrets of social change, I will only refer to the successes or failures of those movements when they might help illuminate some point about successful tactics in social innovation moments. Instead, I will mostly train my focus on efforts to bring about progressive social change. The insights I discuss here might also serve nonprogressive causes; nevertheless, those are not the main focus of this work.

An in-depth look into how the American Legion secured passage of the G.I. Bill uncovers some of these lessons that emerge about this interplay between technology and social change. With the American Legion, the national network it had developed took advantage of the best available technologies at the time to promote and ultimately secure passage of the G.I. Bill. But the Legion's leaders did not just use technology in creative ways. More important than the technology they used to advance this change was what they did with that technology.

First, the supporters of the G.I. Bill used technology to activate a diverse and widespread network of volunteer civic groups that could promote the cause of securing these broad, generous, and inclusive benefits. This network is what is sometimes called "translocal," consisting of small, local chapters tied together in a nation-spanning web that crosses the geographic boundaries of cities and states. Such networks would bind supporters together in a way that not only facilitates cooperation at the local level, but also enables leadership at all levels of the network to activate those local members to engage in advocacy in furtherance of the national cause. In the case of the American Legion, its members met face-to-face in local Legion halls to converse, strategize, and offer mutual aid to one another as they sought to garner support for the effort to pass the G.I. Bill. They also fed information to the leadership about the needs of local returning service members and lobbied their local elected officials, newspaper editors, and community leaders to support the program. This responsive network was essential to the creation of the bonds of trust, support, and cooperation that any grassroots movement needs to bring about social change. What is more, using all manner of communications technologies at their disposal, the leadership of the Legion facilitated the flow of information and the guidance needed to help coordinate efforts organized around a central theme: pass the G.I. Bill. Thus the communications technology available to the Legion made it possible to connect a diverse, nation-spanning network of local cells that helped build trust and mutual support at the grassroots level. The Legion leveraged this translocal network to carry out a broad, national campaign. That translocal network of trust, facilitated by the most advanced communication technologies available at the time, is ultimately what helped the Legion accomplish its goal of passing the G.I. Bill, and it was an indispensable component of that success.

A deeper understanding of the substance of the bill helps uncover an additional core component of the American Legion's strategy and tactics for getting it passed. The Legion did not just build a network and use technology to do so. As a part of this national, grassroots effort, the Legion also rallied support for passage of the benefits program by stressing its near-universal and inclusive components that ignited the passions of avid supporters and sparked broad-based support. The message they conveyed because of these characteristics of the bill was that those involved in the war effort, no matter for how long and whether they saw combat, should have support from the nation they served once they returned from fulfilling their military duties.

The primary goal of the G.I. Bill was to prevent returning service members from falling on economic hard times on their return to civilian life. But it had additional benefits as well. The bill's success would create an engine of economic growth that would transform the nation and lead to a postwar era that saw a steady decline in economic inequality, an increase in civic activism, and the rise of the civil rights movement. The ripple effects of the G.I. Bill were felt for generations and helped transform a nation. "Future historians," as Peter Drucker once said, "may well consider [the passage of the G.I. Bill] the most important event of the twentieth century."[6]

This experience of the American Legion highlights the central insights of this book. Placing tools at the disposal of grassroots groups has helped shape and strengthen efforts designed to promote social change. It has also led to innovation and adaptation as to how those tools can be used and wielded. The first component of successful social change in these social innovation moments appears to be that successful social movements have embraced the most modern means of communication available to them at the time as a tool for organizing and movement building. Indeed, the leaders of and participants in social movements have both adopted and adapted to the tools at their disposal to serve their strategic ends in new and creative ways. Early American movements delivered petitions to Congress seeking all manner of policy reform, just as online petitions are deployed—albeit with lightning speed—today. Pamphleteers urged their supporters to attend rallies and form local chapters of national organizations. In carrying out these tactics, they used printing presses, the mails, and the railroads to fight slavery and child labor, just as activists across the world now use Twitter and Facebook to connect, organize, and mobilize for collective action. Yet social movements are not only about the tools available to such movements to communicate.

In certain, distinct moments, advances in the ability to communicate do appear to correspond to the rise of new social movements. What the American Legion's experience shows, and what the other social movements I will assess in this book also make clear, is that innovations in communications technology can help bring about social change when they are deployed to foster the growth and proliferation of translocal networks: again, small, local groups that are embedded in a broad-based collection of similar organizations. Further, as with the American Legion, movements made up of such networks tend to have success in achieving social change when those translocal groups promote a unifying message that is designed to combat social inequality in one or more of its many forms. The examples of the passage of the G.I. Bill and other similarly successful social movements help illuminate this interplay between technology, social movements, and social inequality. Like the American Legion, groups that do three things well tend to be those that are able to bring about social change.

First, the movements that arose in social innovation moments that I will look at throughout this work harnessed the newest means of communication at their disposal to further their social change goals. In fact, new advances in communications technologies do not just appear to spur social movements. In some ways, they also tend to help shape them: groups in the early nineteenth century mimicked the postal network, and the organizations in the first decades of the twenty-first century are built on the bits and bytes of digital networks.

Second, these groups have often organized translocal, grassroots networks. These networks are made up of "cells," or nodes connected to a larger organization that often spans the nation. This network, rooted in local components but connected to a larger enterprise, encourages face-to-face interactions between participants in the movement. These face-to-face interactions help build trust and feelings of mutual support. The bonds these interactions create help foster cooperation and a willingness to partner with other members of the group to advance social change. Once these local bonds are created, the linkages between small cells that are part of a larger effort facilitate the channeling and coordination of action at the local, state, and national level to bring about social change.

Third, successful social movements in these social innovation moments have often created networks that brought people together from different walks of life and backgrounds through the use of positive, inclusive messages that tapped into the shared interests of the members of the network. Effective social movements promote optimistic and inclusive messages that stress that potential members of a movement share common interests and their destinies are intertwined.

These three components—communications technology, translocal networks, and an optimistic and inclusive message—have been essential elements of social change in critical moments in U.S. history, when technological change and social forces coalesced to create an environment ripe for social change. While this pattern connecting technology, networks, and inclusive messages has emerged, I am quick to admit that one era stands out as a time when individuals attempting to advance social change embarked on a different approach, an approach brought about by an advance in communications technology. This exception may prove the larger rule. It may also help identify not just what was lost as a result of social movements embracing a change in technology, but what also might be gained by avoiding the pitfalls their experiences exposed.

In that era, from the late 1960s through the early 2000s, the social movements that emerged took on the shape of the most advanced communications technology available to them. Then, in the late 1960s, soon after the significant successes of the civil rights movement in achieving landmark legislation that banned racial discrimination in voting, commerce, housing, and employment, something changed in the shape of such movements, thanks to advances in communications technology. Indeed, a new technology appeared, one that broke down the traditional connection between social movements, their ability to communicate, and their translocal character. While many organizations that emerged during this time attempted to combat inequality—gender inequality and inequality based on sexual orientation, for example—such movements utilized the most modern communications technology available to them to create national organizations that had a significantly different structure from those utilized by social movements in the past. This structure mimicked the new communications technology available to them, a new technology that offered them the ability to create a national organization built on little more than a computerized database of names and mailing addresses. This new technology severed the link between communications technologies that fostered the creation of networked, translocal groups and changed not only the shape of social movements but also their focus, enabling them to embrace narrower, less inclusive messages because they could target them toward more discrete, specific populations. Such technology thus changed the shape and focus of social movements and perhaps altered the course of social change itself, at least temporarily.

But just as the computerized mailing lists impacted two generations of advocates, new, powerful communications technologies have, once again, emerged. Over the last decade, there have been some leaps forward, as mobile technologies and web-based organizing have begun to take hold. Have such technologies created a new social innovation moment? If so, can they learn from past social innovation moments to rejuvenate a style of organizing that encourages in-person communications, develops translocal networks, and promotes inclusive messages? I argue here that there is a relationship between technology, networks, and message that helps explain how social change happens in the United States in these social innovation moments and reveals the connection between these three components of change, what I call the *social change matrix.*

A matrix is "something within or from which something else originates, develops, or takes form." In mathematics and science, the term can be depicted in rows and columns or described as an extracellular substance on which other cells are embedded.[7] A matrix of social change is a platform on which different components work together to create social change, with each component connected to and strengthening the others in multiple ways. Activities and communications run in many different directions, reinforcing the efforts carried out through the matrix so that the whole of the activities is greater than the mere sum of the parts. The connection between the three components of the social change matrix is evident throughout the history of social change in the United States at those times when new modes of communication made new ideas and new movements possible. Indeed, at each moment when the ability to communicate improved, social movements have sprouted up, embraced the technology, and used such advances to further their ends. As media theorist Clay Shirky posits: "When we change the way we communicate, we change society."[8] From the Sons of Liberty to the movement for marriage equality, when new technology has made connecting easier, it has shaped the movements that have utilized it, and left its lasting mark on the nation.

To explore these ideas in greater depth, in the first three chapters I will identify the three components of the social change matrix and how they have played out through U.S. history. In chapter 1 I address the first component of the matrix, what I call medium: the mode of communication a group uses to communicate and organize. My review of the advent of the printing press, the post office, the telegraph, the transcontinental railroad, the telephone, the radio, and the television reveals that with the emergence of each of these innovations, a mass movement or movements rose up in their wake. But these advances in communications technology did not cause such movements to take hold and have success when they did. I argue that there are two other critical components to social change success in these social innovation moments.

In chapter 2, I explore the second critical component of the matrix, the network, or the idea that many of the social movements embraced a translocal organizational structure that mimicked the means of communication available to them. This chapter draws from sociology, political science, and political theory to try to understand why some social movements have succeeded in these social innovation moments and why others have failed, with a particular focus on the tactics used and group structures formed by these different movements. I explore whether a movement's structure—which is often a function of the means of communication used by a movement's leaders—might help determine whether that movement can garner the sustained support it needs to bring about successful social change.

In chapter 3, I describe what I identify as the final element of the matrix: that of message. Many of the social movements identified here often embraced a unifying message that sought ways to attract a wide and diverse group of supporters. This chapter utilizes social movement theory to identify what are often seen as the elements of social movement success. It applies that theory to the components of the social change matrix to shed light on the ways that social movements can bring about change in social innovation moments. It also draws on my own experiences as a lawyer and community organizer working for fourteen years in low-income communities in New York City, primarily in Harlem, Washington Heights, the South Bronx, and Chinatown. This work was almost primarily face-to-face and door-to-door, working at the bottom of the organizing pyramid.

While the first three chapters highlight some of the components of successful social change in moments when the means of communications changed, not all organizations have deployed the most advanced communications technologies available to them to build translocal networks and embrace a unifying, broad-based message. Chapter 4 recounts the radical change in communications technology that helped launch many organizations that abandoned the translocal organizing structure because the most modern means of communication available to them—the computerized mailing list—made it easy for them to do so. I explore some of the ramifications of this shift in organizational structure and the effects it had on progressive social change.

Following this review of past social innovation moments, starting with chapter 5, I begin to describe the social innovation moment we are in now. It sets the stage for the next chapters by analyzing a contemporary social mobilization effort: the West Virginia teachers' strike of 2018. This chapter will assess that effort in light of the capacities that new tools—like the Internet, social media, and mobile technologies—offer for building and strengthening movements. In this chapter and the ones that follow, I borrow from contemporary research on the potential impact of digital communications tools on organizing but also draw from new, primary, and qualitative research of contemporary social movements at work. Specifically, chapters 6 and 7 recount several recent campaigns to advance social change to examine their success at deploying communications technology, building a network, and crafting an inclusive message, including the effort to reauthorize and expand the reach of the Violence Against Women Act and the campaign for marriage equality, with particular emphasis on an effort to pass a pro–marriage equality ballot referendum in Maine in 2012. Chapter 8 describes a campaign to raise the minimum wage for hotel workers in Long Beach, California, exploring the deft use of the social change matrix to address income inequality in a way that was adapted to local conditions.

In chapter 9 I offer some concluding thoughts on these subjects, showing that lasting social change must come from networked activists whose efforts are boosted and amplified by technology and who promote an inclusive and positive message. I also address some of the downsides of digital technologies and the potential threats they pose to social movement success.

It is my hope that this book will help readers understand the role of technological and social innovation in those processes that bring about social change. In many ways, this effort to understand the interplay between social movements and communications technology is a daunting task, one that I try to approach with humility. I do not anticipate that this book is the last word on these subjects. Rather, I hope to begin a dialogue that helps the reader see some of the patterns exposed here so that we can make sense of this connection together, learn from the past, and harness the social innovation moment I believe we are in today. In this book, I try to follow what appear to be some consistent themes in the history, research, and lived experience of social movements in particular moments in time so that we might learn from their successes, and their failures.

One of my central themes is that the creation of civic connection and civic innovation is deeply tied to the means of communications available to leaders, organizers, and average citizens. The historical record connecting technological and social innovation to social movements is rich, deep, and long, and offers hope that the innovations of today can foster the type of civic movements that can bring about social change that is broad-based, widely supported, and desperately needed—provided we learn the lessons of successful social movements of the past. This book explores these successes and strives to help advocates understand how to harness the elements of the social change matrix to bring about lasting, effective, and real social change at a time when many are looking for guidance on how to do so. I hope this book offers a glimpse into the world of the possible in our present social innovation moment and at the critical moments that will inevitably arise in the future.

1

MEDIUM

John Winston Jones probably feared the future. And he probably feared change too. A lawyer from Virginia elected to the House of Representatives as an Andrew Jackson Democrat in the mid-1830s, he would rise to prominence and become Speaker of the House for one term in the mid-1840s. Speaking in opposition to the passage of legislation to abolish slavery in the District of Columbia, he paid particularly close attention to the ways that new technology and the mails strengthened the work of abolitionist societies. Jones lamented the emergence of these groups, fueled by the new technology of printing presses "worked by steam" as he would say. For Jones, these groups had "the open and avowed object of effecting the immediate abolition of slavery in the southern States." The abolitionists used the postal system and the new steam printing press, what Jones called the "great revolutionizers of the world," to "circulate . . . throughout the slaveholding States, large numbers of newspapers, pamphlets, tracts, and pictures, calculated, in an imminent degree, to rouse and inflame the passions of the slaves against their masters, to urge them on to deeds of death, and to involve them in all the horrors of a servile war, productions which . . . were as foul libels upon the people of the South as were ever printed."[1]

Communications technology, in the form of the steam printing press, combined with the reach of the postal system, helped spur abolitionist efforts. With the application of steam power to printing technology in the 1830s, the cost of printing a newspaper was reduced from six cents to one. Where traditional presses could print about one thousand impressions per hour, the new steam presses were capable of five times that many.[2] Couple this innovation with the reach of the postal system to practically every community in the United States and the abolitionist movement had both a means and a method for spreading its antislavery message throughout the nation. Mass production of abolitionist literature ensued and shipments of the tracts were distributed throughout the country by the mails.

With the introduction of a new means of communicating through the invention of the steam-powered printing press, there emerged a new social movement. Indeed, just as the abolitionist movement was gaining strength, this new technology helped fuel the advocacy of the movement and strengthen its power and reach. The steam printing press, and the postal system abolitionists would use to spread their message, became critical means of advancing social change. By combining new technology with a nation-spanning lattice of the postal system, a network of abolitionist groups was able to spread the word about the antislavery cause and rally supporters to it. In these ways, the abolitionist movement itself was bound up with the technology that facilitated its emergence, extended its reach, and strengthened its impact. This connection is not accidental, and it is deeply engrained in U.S. history. This chapter explores this connection, from the events leading up to the American Revolution through the successes of the civil rights movement.

The American Revolution and the Printing Press

As in all of the social movements I will recount here, the history of the revolutionaries in colonial America was tied inextricably to the ability of the colonists to communicate with one another. Prior to the military exchanges of the spring of 1775, the ability to communicate would become a central flashpoint in the events leading up to the revolution. Indeed, the Stamp Act crisis of the 1760s in many ways set the stage for the upheaval to come. As we will see in other points throughout U.S. history, the most important technology at the time—the printing press—would become the hub around which social change evolved, both in the events leading up to the revolution and during the war. Even before the Boston Tea Party, colonists banded together—often spurred by operators of printing presses—to resist the passage and implementation of the Stamp Act in the 1760s. Struggling with debts incurred waging the French and Indian War, the British Crown imposed a tax on paper used to print everything from newspapers to contracts. This had a direct impact on the printing industry and affected printers' profits, as newspapers and other printed matter became costly to print and more expensive to purchase. Printers, concerned for their livelihoods, criticized the act as a symbol of British tyranny. They sowed the seeds of a refrain—"No taxation without representation"—that would become a rallying cry a decade later. In a widely publicized act, the Virginia legislature, one of the first governmental bodies to take action against the new legislation, passed the "Virginia Resolves," which declared that "the Taxation of the People by Themselves, or by Persons Chosen by Themselves to Represent them . . . is the only Security against a Burthensome Taxation."[3]

The British Parliament may have made a critical error in targeting the printed word for taxation. Printers had become central to the formation of a nascent national identity in the mid-1700s.[4] To target their livelihood—the paper on which they printed their newspapers, the legal documents people used, even law licenses—drove up their cost of doing business and threatened their ability to earn a living. As a result, printers, the individuals most responsible for fostering communication among the colonies, became the object of the tax. It is no surprise that they were some of the most vocal opponents of it and used the most effective means at their disposal—the power of their presses—to whip up colonial sentiment against the law. Even before the act went into effect in late 1765, a new movement, the Sons of Liberty, a group that would play a critical role in the coming decade in terms of rousing colonial sentiment against the Crown, arose with the express purpose of protesting the imposition of the tax.[5] Opposition to the act was so strong in the colonies that Parliament was given little choice but to repeal it. Already strapped for funds, the prospect of raising a large army to enforce the law, which would have been necessary given colonial opposition, seemed daunting and likely to further stoke rebellious sentiment across the Atlantic.[6] In February 1766, Parliament ultimately repealed the act, just months after it had gone into effect.

While the Stamp Act crisis foreshadowed the events that would unfold in the next decade, the ability of the upstart revolutionaries to communicate became an important issue to address even before the first real battles broke out. When the British controlled the official postal system, the mails were neither protected nor private. British officials could read the letters and newspaper tracts they were asked to transport through the official postal system. Because of the threat to the privacy of their correspondence, many rebel colonists feared charges of sedition should they express sentiments, in writing, in favor of independence. In response, private messengers often transported communications outside of the official postal system. Agitators in Boston, at the urging of a newspaper publisher from Baltimore who had started a private route between that city and Philadelphia, began to advocate for the creation of what was to be called a "constitutional" postal system that would operate independently of the official British one.[7] Those who favored a separate system argued that the costs the official system imposed represented an unfair tax that lined the coffers of the British government, that loyalist postal clerks could monitor and stifle proindependence communications, and that an independent postal system would foster a more cohesive colonial body.[8]

For the early proponents of an independent, intracolonial postal system, such a network was essential to uniting the colonies in support of the growing rebellion.[9] Before it even voted for independence, the Continental Congress met in Philadelphia in the summer of 1775 and created what it called the Constitutional Post, which had routes that stretched from Maine to Georgia, naming Benjamin Franklin as the first postmaster general.[10] After repeal of the Stamp Act, many printers continued to stoke sentiments in favor of revolution, while others vehemently supported loyalty to England. Because they served as both mouthpieces and symbols, of either revolutionary or Tory sympathies, printers were often singled out for violence. When British forces occupied Boston in 1774, "an enemies list" of sorts was drawn up and circulated throughout the army. It identified those revolutionaries to be put "to the sword" should rebellion arise, including two local printers, who were referred to as "trumpeters of sedition."[11] On the American side, mobs would sometimes resort to violence to silence the loyalist press, descending on printing presses to scuttle them and threatening—and sometimes inflicting—bodily harm on the loyalist publishers.[12] Even the iconic date of July 4, 1776, symbolized the importance that the founding fathers placed on their ability to communicate as well as the centrality of an effective means of communication. Indeed, the event that took place on the fourth was not the signing of the Declaration of Independence as popular perception holds. No, the Fourth of July marks the date a final version of the document was approved by the Continental Congress and transmitted to the Philadelphia printer John Dunlap.[13] In other words, what we celebrate on this day in U.S. history is when the founders pressed "send."

Printers supported the rebellion by distributing their pamphlets throughout the colonies, building and sustaining support for independence. As one Tory lamented, "The peasants and their housewives in every part of the land were able to dispute on politics and positively to determine upon our liberties."[14] Printers spread their tracts throughout the country to other printers. They would regurgitate news from other parts of the colonies in their own publications, which helped keep the people informed of important events and developments unfolding elsewhere.[15] In these ways, printers, printing presses, and the postal system that carried their revolutionary tracts became essential elements of the great social movement that sparked revolution and independence.[16]

The colonists clearly placed great importance on protecting the means of communication, both by defending printers during the Stamp Act crisis and creating an independent postal system. This should come as no surprise given the critical role the press ultimately played in supporting the revolutionary forces and sympathies and sustaining freedom's fervor through the military hostilities. As David Ramsay, a historian of the revolution, wrote in 1789: "In establishing American independence, the pen and the press had merit equal to that of the sword." For Ramsay, because "the war was the people's war, and was carried on without funds, the exertions of the army would have been insufficient to effect the revolution, unless the great body of the people had been prepared for it, and also kept in a great disposition to oppose Great Britain." Printers were essential to that effort. "To rouse and unite the inhabitants, and to persuade them to patience for several years, under present sufferings, with the hope of obtaining remote advantages for their posterity, was a work of difficulty," he wrote. Maintaining that support "was effected in a great measure by the tongues and pens of the well informed citizens, and on it depended the success of military operations."[17]

The experience of the colonists and their leaders in using the printing press and the mails to further the cause of independence reveals the first insight into the relationship between innovations in communications and social movements. Without the ability to communicate, the colonists would not have been able to band together against a common foe and could not stay informed of developments, maintain discipline and morale, and coordinate military and political affairs. Simply put, without printers and the nascent postal system and communications network they utilized, revolution and, ultimately, independence would have been impossible. What makes this moment in history particularly important as we look at the interplay between advances in the means of communication and social movements generally is that the rapid expansion of the printing press in the colonies in the mid-eighteenth century, together with the creation of the independent postal system, are likely what helped create the environment in which revolutionary fervor could flow in the first great social innovation moment of the American experience.[18] This pattern will repeat itself throughout the history of social movements in the United States. An effective means of communication is essential to any social movement because it is through coordinated community action that any social movement takes hold and makes change; that coordination is facilitated and made possible by the means of communication a social movement uses and has at its disposal. But there are other insights as well revealed by the connection between communications technology and social change in these social innovation moments. The advent of the *post*colonial postal system helps reveal more of those insights.

A New Postal System and New Social Movements

Just as printers and the continental postal system were vital to waging a revolution, the early Congress knew the importance of a vibrant postal system in the new republic. Indeed, the postal system played a critical role as the nerve system of the new nation. The typical post office, located in a local merchant's storefront, was not just a place to pick up one's mail. It became a font of news and gossip and a place where individuals could discuss local issues and national developments. According to the historian Joseph Adelman, the post office, in cities and small towns throughout the country, became a place where people gathered to receive the news and debate it. It was a translocal network that connected small communities to what was happening on the national level. It formed, in Adelman's words, the "crucial circuitry that could facilitate debate that transcended local and face-to-face interactions" and "linked towns and regions to one another in a spatially extended public sphere."[19]

The size and scope of the postal system increased dramatically from colonial times to the 1820s. On the eve of the Revolutionary War, there were just sixty-seven post offices, or four for every hundred thousand inhabitants.[20] By 1828, there were seventy-four for every hundred thousand. The American system had twice as many offices as Great Britain's, and five times as many as that of France (which had four for every hundred thousand residents).[21] By 1831, the American system had more than 8,700 postmasters and employed just over three-quarters of the entire federal civilian workforce. By comparison, the federal army at the time was made up of 6,332 soldiers.[22] According to the historian Richard John, the postal service was so vast, and reached so many areas of the country, that "it would hardly be an exaggeration to suggest that for the vast majority of Americans the postal system *was* the central government."[23] By giving Americans the ability to communicate across the expanse of the nation, and to stay attuned to the developments at the national level, it became the backbone of communication throughout the country. The postal system did not just spread information, however. It also led to a rich dialogue and a broader appreciation for the interests and needs of individuals from different walks of life and communities from different regions in the country. As the Supreme Court justice Joseph Story, in his commentaries on the U.S. Constitution, would state, the postal system "brings the most distant places and persons, as it were, in contact with each other; and thus softens the anxieties, increases the enjoyments, and cheers the solitude of millions of hearts."[24] What the printing press and the postal system also did was spark the emergence of new, nation-spanning, civic associations that held their roots in local communities. What is more, these civic associations looked a lot like the postal system itself. While the postal system may have mapped the growth of societies across the new nation, it also formed the spine on which civic society would form and flourish.

Several contemporary observers of civic life in the early days of the United States help sketch the role, growth, and importance of civic associations during this period. Of course, one of the best places to start when looking at the role of civil society in that era is the writings of the French aristocrat Alexis de Tocqueville, who toured the United States in the 1830s, ostensibly to visit the nation's prison and jail system. His writings paint an illuminating picture of a wide range of topics, from the role of freedom of the press to the conditions of slaves. Perhaps one of his most poignant observations reveals the important functions, breadth, and popularity of civic associations in American life at the time. To Tocqueville, Americans of all backgrounds practiced what he called the "infinite art"[25] of joining together to form civic groups. Such civic groups often included members from different classes and backgrounds (save slaves, of course). "Americans of all ages, all conditions, all minds constantly unite," he wrote, forming groups that were not just commercial in nature. They would also form groups of "a thousand other kinds: religious, moral, grave, futile, very general and very particular, immense and very small." These groups were organized to "give fêtes, to found seminaries, to build inns, to raise churches, to distribute books, to send missionaries to the antipodes; in this manner they create hospitals, prisons, schools." For Americans, he wrote, "if it is a question of bringing to light a truth or developing a sentiment with the support of a great example, they associate." Comparing practices in the United States to Europe, he found that "everywhere that, at the head of a new undertaking, you see the government in France and a great lord in England, count on it that you will perceive an association in the United States."[26] As Tocqueville observed, these groups were formed to tackle problems small and large, local and national: "An obstacle comes up on the public highway, passage is interrupted, traffic stops; neighbors immediately establish themselves in a deliberating body; from this improvised assembly will issue an executive power that will remedy the ill—before the idea of an authority preexisting that of those interested has presented itself to anyone's imagination."[27] Research on the presence of associations in the new nation backs up Tocqueville's observation and reveals that the period from the 1790s through the 1830s saw vast growth in civic life and organizations. For example, civic associations other than formal church groups proliferated in places like Massachusetts. Within the city of Boston, from before 1760 to 1830, there was an increase from thirteen such groups to 135. In the rest of the state, the numbers increased even more, from twenty-four to 1,305.[28]

Another observer of civil society in the first part of the nineteenth century, the Unitarian preacher William Channing, described the "energy with which the principle of combination, or of action by joint forces, by associated numbers," as "one of the most remarkable circumstances or features of our age." Channing would echo Tocqueville when he said that "every thing is done now by societies. . . . You can scarcely name an object for which some institution has not been formed." Channing confirmed Tocqueville's observation that Americans seemed to form associations to take on all manner of tasks and issues: "Would men spread one set of opinions, or crush another? They make a society. Would they improve the penal code, or relieve poor debtors? They make societies. Would they encourage agriculture, or manufactures, or science? They make societies. Would one class encourage horse-racing, and another discourage travelling on Sunday? They form societies." Channing observed further that Americans at the time had "immense institutions spreading over the country, combining hosts for particular objects."[29] While local issues were often the focus of local associations, as Channing recognized, broader movements began to take hold to address a range of national matters. Movements to promote the abolition of slavery, temperance, closing businesses and the post office on Sundays, and women's suffrage all took root in the decades leading up to the Civil War.

Channing saw the postal system and the press, and other technological and social innovations like the steamboat and improved roads, as facilitating intragroup communications. The ability to communicate, coordinate, and organize groups, associations, and nation-spanning networks was brought about by what Channing called "modern improvements": that is, "the post-office, . . . the steamboat, and especially by the press,—by newspapers, periodicals, tracts, and other publications." For Channing it was "through these means, men of one mind, through a whole country, easily understand one another, and easily act together." Because of these means of communication and travel, and the coordinated action they facilitated, "when a few leaders have agreed on an object, an impulse may be given in a month to the whole country, whole states may be deluged with tracts and other publications, and a voice like that of many waters, be called forth from immense and widely separated multitudes."[30]

The private groups that emerged in the first half of the nineteenth century did not just form in the wake of the expansion of the U.S. postal system, or even because of it. No, these social movements, in their structure and reach, would mimic the very means of communication—the postal system—that helped them form, spread, and expand. Social movements formed as national networks of local organizations in structures that replicated the postal system at the time. This idea is central to the movements that emerged in the early nineteenth century and the next 150 years of the American experience, as we will see in subsequent chapters.

These organizations were what I described in the introduction as "translocal." That is, they were small, local chapters united in a federated system, a system that did not just look like the postal system but tracked it, with local chapters forming as the post office expanded. These local units could link to larger efforts and could target their advocacy on different fronts: i.e., their elected officials that served on the local level, but also those that were sent to Washington. Many groups set up these networked structures, had elections to choose representatives at the local and national levels, and had constitutions that embodied the organizations' rules and procedures, just like the federal government itself.[31]

As migrants moved within the country, they often communicated with those from their point of origin to set up local chapters of organizations of which they had been members previously.[32] Organizations seeking to expand their reach mailed printed material to new communities in the hopes of kick-starting new local chapters in those communities. The favorable treatment printed material received in the postal system helped facilitate these communications.[33] At a time of great social innovation—the expansion of the postal system—civic life flowered throughout the entire United States, in no small part because of this strengthening of the means of communication. What is more, while these organizations might have looked like the federal government, given that local chapters formed around post offices, it is probably more accurate to say that they mimicked the postal system, in both structure and reach.

The postal system also helped spark social networks and, ultimately, social movements. Thanks in part to the postal system, the late eighteenth and early nineteenth centuries saw the growth of national movements that were linked to the physical spread of post offices and postal routes. While the Freemasons started before the revolution, other large national groups got their foothold in the period between the early part of the nineteenth century and the beginning of the Civil War, like the Independent Order of Odd Fellows, the American Temperance Society, the General Union for Promoting Observance of the Christian Sabbath, and the Young Men's Christian Association. At some point during their existence, each of these groups accumulated a membership base that included at least 1 percent of the U.S. adult population of either men or women, depending on whether they were single-sex organizations.[34] This development—that social movements seem to follow innovations in the means of communications—is the second insight or lesson that emerges from this analysis of the relationship between social movements and innovations in the means of communications; in fact, this phenomenon represents the idea of the social innovation moment itself. It is revealed in not just the example of the postal system but also the steam printing press that opened this chapter: new social movements often arise after the introduction of a new means of communication. Just as the American Revolution followed the spread and impact of the printing press, many grassroots networks arose in the wake of the expansion of the U.S. postal system. This is a pattern that appears to repeat itself throughout U.S. history, creating social innovation moments. Indeed, innovations in the ways that individuals and groups could communicate with each other in the early nineteenth century—namely, the expansion, sophistication, and spread of the postal system—created an environment that spurred the growth of these national networks of local groups.

Seneca Falls, the Telegraph, and the Birth of the Women's Movement

Even as abolitionists used the printing press and the post office to spread their antislavery tracts, a new technology emerged that would make real-time communication possible. Short of smoke signals and semaphores, never before had communications technology been able to truly collapse time and distance, as the telegraph would. This new mode of communication would transform business communications, the newspaper industry, and government operations. It would also help spread the word of a new movement, one that promoted equal rights for women.

Women active in the abolitionist movement turned their sights on the rights afforded half of the population. Lucretia Mott and Elizabeth Cady Stanton, who would become leaders in the women's movement in the United States, met in 1840 in London at the World Anti-Slavery Convention. Women could not serve as delegates at the convention, however, and Mott was refused the opportunity to speak formally there.[35] Stanton's memoir explains the reaction to the treatment of women at the convention as follows: "As the convention adjourned, the remark was heard on all sides, 'It is about time some demand was made for new liberties for women.' As Mrs. Mott and I walked home, arm in arm, commenting on the incidents of the day, we resolved to hold a convention as soon as we returned home, and form a society to advocate the rights of women."[36]

In 1845, the ability to communicate changed dramatically when Samuel Morse sent his first telegraph message. By 1848, the newspaper industry adapted to this new means of communication, which relayed news across the country in real time. In May of 1848, six New York newspapers banded together to form the Associated Press, which allowed them to pool resources and take advantage of this new technology.

In the summer of 1848, Mott and Stanton called for a women's "convention" in Seneca Falls, New York. They sent word out to local newspapers to advertise the convention, and roughly three hundred women and men attended on the relatively short notice of less than two weeks. The two-day convention was attended by Mott and Stanton and such leading abolitionists as Frederick Douglass. News of the convention was reported in papers throughout the country, transmitted by the telegraph. While other efforts to advocate women's rights had certainly been made in the past, according to the historian Daniel Walker Howe, the publicity that was spread through the newspapers about the Seneca Falls Convention, made possible by the telegraph, is what enabled these proceedings to garner the attention of women and men across the country.[37] Soon thereafter, advocates held more such conventions and meetings, launching the women's movement, fueled by the communications technology that could spread the word about the cause.

For some women, the fact that the telegraph united the world through the possibility of real-time communication meant a new dawn was emerging, one in which stronger rights for women might bloom.[38] One leader of the women's movement, Emma Coe, in comments closing a women's rights convention in Worcester, Massachusetts, in 1851, made this connection explicit, proclaiming the beginning of a "new era," one that was "heralded by a thousand harbingers," including the "lightning coursing the telegraphic wires" as well as "the smoke-girt steeds rushing along our iron-rimmed ruts." These were "but embassies of a power whose will will yet place freedom upon something more than a theoretical basis, and give equality of privileges a being as well as a name."[39]

Technology and the Progressive Era

The fight over slavery reached its boiling point with the Civil War. But other developments were also afoot in the middle of the nineteenth century, technological changes that would have a profound effect on the ability to communicate, associate, travel, and organize. Steam-powered boats and trains moved passengers more quickly and to farther reaches of the nation, and the introduction of the telegraph helped bend time and space and facilitated long-distance, instantaneous communication for the first time in history. Experiences uniting for a common cause during the Civil War whet American appetites for collective endeavors and engaged organizations, particularly in the North, where victory gave many the opportunity to work together successfully to pursue and achieve common goals. As Harvard's Theda Skocpol explains, the Civil War had a significant impact on civic life in the decades following the armed struggle, and the shape of that civic engagement would replicate the newly strengthened national government: "People committed themselves to service; and massive wartime efforts reinforced the practicality of popularly rooted federalism as America's preeminent model for large-scale association building."[40] This was especially true for leaders, she says, for whom "wartime experiences created ideals, network connections, and models of citizen organization that encouraged ambitious association building long after the fighting ceased."[41]

While the war and its aftermath certainly had a drastic impact on social life, rapid technological and social innovation would soon create further social change throughout the United States. The introduction of the telephone, further expansion of the railroad, innovations in manufacturing, innovations in the organization of businesses and factories: all of these would have profound impacts on community life at the turn of the twentieth century. Although government grew in size during this period, the main source of support and solidarity in individuals' daily lives were the civic associations to which they belonged: local business groups, trade unions, farmers' associations, mutual aid societies, and women's leagues. A growing population, a powerful business elite, and a rapidly expanding economy were all shedding the vestiges of their agrarian, local roots and entering the age of cities and factories; national corporations; railroads and movie theaters; the *Saturday Evening Post* and the Sears catalog; automobiles, trains, steamships and, ultimately, airplanes.

In this period, what came to be known as the Progressive Era, translocal organizations thrived, creating networks that agitated for greater social equality. Trade unions expanded, with the American Federation of Labor growing in ranks from 150,000 in 1896 to 500,000 in 1897.[42] By 1917, 3 million workers, or 11 percent of the nonagricultural workforce, were unionized.[43] Starting in the mid-nineteenth century, new women's groups formed and women's colleges were established.[44] Organizations advocating for the rights of immigrants and African Americans also arose. This era saw the formation of the National Association for the Advancement of Colored People (NAACP), which boasted 165 local chapters by 1918,[45] and the National Association of Colored Women.[46] Groups of black farmers formed their own associations when they were excluded from those made up of white farmers, like the Colored Farmers' National Alliance.[47] Ethnic groups of recent immigrants formed organizations to help their constituents adjust to life in the New World.[48] Organizations like the Young Men's Christian Association and the Salvation Army worked to alleviate financial distress, particularly in urban centers.[49] Just as in the early part of the nineteenth century, translocal groups flourished, as participants tried to join together to have an impact on the forces changing their lives and to restore a sense of community.

This explosion in civic life corresponded to advances in the ability to communicate and travel, which means that this period is easily identified as a social innovation moment. During this period, perhaps like in no era before it, technological and social innovations were at the intersection of civic engagement. Indeed, civic life thrived, as members in networks of translocal organizations could communicate with each other more easily; travel between communities more rapidly; send delegates to regional and national conventions with relative ease; and harness the power of their collective strength at the local, state, and national levels. Frederick Harrion, a British visitor to the United States in 1901, wrote: "Life in the states is one perpetual whirl of telephones, telegrams, phonographs, electric bells, motors, lifts, and automatic instruments."[50] The products of Bell, Edison, and countless other inventors and tinkerers transformed life and connected people in ways only dreamed of before, ushering in the era of the telephone, the electric light bulb, the assembly line, the railroad, and the motion picture.[51] Individuals found new ways to work together, play together, communicate, and engage in commerce. Technological and social innovation brought businesses, communities, and individuals together like never before.

Technological innovation in the areas of communication and travel perhaps made the greatest impact on the ability of individuals, businesses, and communities to communicate over great distances and facilitated easier and more rapid in-person communication as well. Long-distance communication was made easier with the invention of the telephone. While the telegraph had linked the nation for nearly half a century, the introduction of the telephone brought the ability to communicate instantaneously and without the aid of the telegraph clerk. By 1910, 1 million homes had telephones,[52] aiding the communication of individuals and the ability of business managers to communicate with the outposts of widespread business networks. Apart from this direct, though long-distance, means of communication, the ability to travel great distances was made much more convenient with the rapid expansion of the railroad. By 1893, four transcontinental railroads traversed the United States.[53] Transcontinental travel, through the rails and later the automobile, and transcontinental communication, from the telegraph and telephone, connected the nation from coast to coast like never before.[54]

These advances in technology aided and fostered the growth of civic associations and the richness of civic life. Local chapters could communicate directly with those in leadership positions, and national leaders could travel more easily to conventions, marches, rallies, and speeches. National figures like William Jennings Bryan would traverse the nation over train lines, often sleeping in train cars as he rumbled over the rails between towns en route to his next speech.[55] Suffragettes in upstate New York organized a "trolley car campaign" in which they held rallies in the cities between Syracuse and Albany connected by an intercity rail line.[56]

As Skocpol points out, even from the early days of the railroads, civic leaders took to the rails to promote the missions of their organizations. She recounts the work of Frances Willard of the Woman's Christian Temperance Union, who "visited every U.S. city of five thousand people or more at least once during the

1870s and 1880s and was therefore always on the train and hardly ever 'at home' in Evanston, Illinois."[57] The task of recruitment of new members and retention of existing members fell to national leaders who would use transcontinental travel to move among their local chapters. For Skocpol, this sort of travel was critical as she describes the ethos among leaders of the era as "interact or die."[58]

The innovations of the telephone and the railroad helped either launch or expand the translocal networks that thrived during the Progressive Era, whether they were formal, national, Progressive Movement organizations; modern media outlets; or megacorporations. They all expanded together, creating a new national and cultural paradigm: nation-spanning entities deeply tied to the nation-spanning technologies. Thus this social innovation moment led to changes in all aspects of civic life. In government, business, and civil society, new methods of communication and travel helped strengthen existing organizations and spurred the creation of new organizations and entities, launching campaigns and practices that benefited from the ability to have such national reach and impact. New and established organizations benefited from these advances in communications technology and innovations in travel that facilitated both face-to-face communications and community organizing. Such organizations also mimicked these technological and social innovations in structure and reach. Just as railroads spanned the nation in translocal networks, national organizations followed that same pattern. They were translocal entities that had local centers of membership embedded within a nation-spanning network structure. They were, to borrow again from Skocpol, "federated" entities, designed to "span the nation and tie localities and states together."[59] And they would bring about significant changes to the way society worked in the early part of the twentieth century, from securing the right to vote for women, to the adoption of workplace and food safety measures, to the institution of a federal income tax and the direct election of U.S. senators. Much of this was accomplished through the painstaking work of direct, face-to-face organizing that occurred within nation-spanning, human networks made more effective through the advances in communication and mobility that occurred in this era.

The Radio, Roosevelt, and the New Deal

While previous periods of civic activism seem to have been preceded by the emergence of a new means of communication, during the Great Depression, as a new means of communication emerged on the scene, civic life was less robust.[60] But just as the postal system, telegraph, telephone, and transcontinental railroad facilitated communication and cooperation, the new technology of the radio aided President Roosevelt in his campaign to garner support for his New Deal programs and rally the nation behind them. His homespun style resonated with dirt-poor farmers and immigrant factory workers alike, who could not have been further in economic stature from the blueblood patrician from Hyde Park, the Groton School, and Harvard. Yet Roosevelt was able to harness technology to communicate a message of harmony and collective can-do in a way that fostered trust and engendered support for the social programs of the New Deal. What Roosevelt's deft use of the radio reveals is the role that communications technologies can play in fostering trust, building personal connections, and linking otherwise differing interests together to address large societal challenges. Analysis of the role of the radio in Roosevelt's efforts anticipates some of the conversations on these topics to come in subsequent chapters.

While radio had been invented decades earlier, at the beginning of the 1930s, only 40 percent of the population had access to a radio in the home; by decade's end, that number reached nearly 90 percent.[61] While the forces beyond the reach of most Americans, both nationally and abroad, may have unmoored many from traditional relationships and a sense of control over their lives, the radio helped make sense of those changes and connected people to these greater forces. Although radio broadcasts could not substitute for personal interactions, listeners developed intimate feelings for the characters that would enter their living rooms through the radio. They also developed a common vocabulary through which they could build closer bonds with their neighbors, friends, and coworkers by discussing the lives of fictional characters in radio serials and political developments in Washington and across the globe.[62]

In 1933, amid the darkest days of the Great Depression, Roosevelt took office and used the power of radio to instill confidence in those of his policies designed to address the economic situation and restore trust in the financial system. Immediately after reciting the oath of office, he took to the radio to deliver his first inaugural address from the steps of the Capitol. Filled with martial rhetoric, his speech attempted to instill a sense of confidence in his new administration.[63] Of course, it was in this address that Roosevelt uttered these immortal lines: "So, first of all, let me assert my firm belief that the only thing we have to fear is fear itself—nameless, unreasoning, unjustified terror which paralyzes needed efforts to convert retreat into advance."[64] Public response to the address was overwhelmingly positive, and Roosevelt launched his initial efforts to address the economic situation, including imposing the four-day "bank holiday" to shore up the nation's financial system, which was in free fall, suffering from a near fatal collapse in confidence.[65]

Roosevelt believed radio could "restore direct contact between the masses and their chosen leaders."[66] Roosevelt's radio address on March 12, 1933, the first of his famous "fireside chats," starts out with language that is straightforward and clear, designed to draw the audience in and convey to them that the president understood what Americans had been experiencing with the hardships imposed by the bank holiday. He also set out to explain to the people of the United States, in simple terms, why he did what he did and what he intended to do in the future. "I want to talk for a few minutes with the people of the United States about banking," he began, "with the comparatively few who understand the mechanics of banking but more particularly with the overwhelming majority who use banks for the making of deposits and the drawing of checks. I want to tell you what has been done in the last few days, why it was done, and what the next steps are going to be." Roosevelt expressed his appreciation for the shared sacrifice made by citizens as a result of the bank holiday and he stressed the need to ensure that the population was aware of the reasons they were facing such hardships as a result of the government's actions with respect to the banks. He stressed that the efforts of the federal and state governments to regulate the banking industry, although "couched for the most part in banking and legal terms should be explained for the benefit of the average citizen." He noted "the fortitude and good temper with which everybody has accepted the inconvenience and hardships of the banking holiday." Roosevelt made the following appeal: "After all there is an element in the readjustment of our financial system more important than currency, more important than gold, and that is the confidence of the people." He concluded with a line designed to rally the nation and bind everyone together in the task at hand: "It is your problem no less than it is mine. Together we cannot fail."[67]

The speech worked. Almost instantly, many Americans stopped withdrawing their funds from the banks and, instead, began making deposits.[68] Ninety percent of the banks closed during the bank holiday were able to reopen.[69] The *New York Times* commented that "the President's use of the radio for [the purpose of restoring faith in the banking sector] is a fresh demonstration of the wonderful power of appeal to the people which science has placed in his hands." It stressed the power the radio gave Roosevelt to enter the homes of the citizenry and develop relationships with those listening: "When millions of listeners can hear the President speak to them, as it were, directly in their own homes, we get a new meaning for the old phrase about a public man 'going to the country.'"[70] Comparing Roosevelt's newfound power with that of President Wilson just a decade before, the *Times* stressed that when the latter attempted such direct communications with his constituents, "it meant wearisome travel and many speeches to different audiences." In contrast, Roosevelt "can sit at ease in his own study and be sure of a multitude of hearers beyond the dreams of the old-style campaigner."[71]

While Roosevelt was building support for his policies through the radio, the severe Depression impacted civic participation. While national organizations enjoyed growth from the early part of the twentieth century up to 1930, the lost decade of the Depression years saw a marked decline in membership in the large national organizations that had thrived at the turn of the century.[72] According to one individual who wrote Roosevelt, this may have been in part a function of the very fact that the radio may have displaced the personal need people felt for face-to-face communications and local civic engagement:

> You took advantage of the radio and the great American Sunday evening at home, and for the second time talked over the problems of the day as a family matter with the families of the nation. You talked as easily and as informally as a neighbor who had just dropped in to visit the folks. There was no more authority, mystery or pose about your talk than in the old time political arguments around the stove and cracker barrel of a country store or in the old-fashioned wooden Indian city cigar store. Truly Mr. Roosevelt you revived the modes and manners of the primitive forums of American democracy. The old town meeting is now a nation's meeting.[73]

It is possible that the radio may have played some role in diminishing the importance of grassroots groups and community-based, face-to-face organizing. Yet in the wake of the Depression, during the subsequent war years, we see that at least some translocal organizing was still occurring, as the success of the American Legion in promoting the G.I. Bill proves. What is more, the success of the American Legion and other groups that advocated for the G.I. Bill would lead to the next great social movement to arise in a social innovation moment: the effort to promote civil rights for African Americans.

Technology and the Civil Rights Movement

Returning service members did not just seek and obtain benefits through the G.I. Bill. For some, their experience fighting totalitarianism, anti-Semitism, and racism abroad ignited a desire to bring the cause of racial equality home. While African American service members benefited from the G.I. Bill, it was not in the same way that white service members did. The black service members were returning to a country still ravaged by racial discrimination: in educational institutions, private accommodations, housing, and employment. These veterans and the countless youth and women who came together to fight the Jim Crow system in all its forms would change the course of U.S. history. It turns out it would also be the last gasp of true networked, grassroots movements, at least for some time.

The period after World War II, perhaps as a result of the benefits offered through the G.I. Bill, was considered a "golden age" of civic participation. Yet while civic groups of women and people of color certainly arose during this time, segregation—formal political and economic discrimination—was the order of the day. But this period was also marked by the rise of the civil rights movement, which worked to lessen racial inequality. As in past eras, the rise of the civil rights movement corresponded to the introduction of a new and powerful technology, in this case one that beamed images of the movement's struggles throughout the world: television. While grassroots organizing and legal tactics were both deployed to further the cause of the movement, television played a critical role in raising awareness about the plight of black America and to garner sympathy and support for the movement's demands.

For many returning service members, like Medgar Evers, their experience fighting for democracy in World War II led them to question and challenge the segregation they faced in the Jim Crow South. In 1954, the Supreme Court's decision in *Brown v. Board of Education*, which struck down racial segregation in public schools, helped fuel advocacy efforts and raised awareness about the evils of segregation. During the height of the Cold War, the treatment of African Americans by the government and private businesses became a point of tension for the United States on the global stage, as the Soviet Union battled for the hearts and minds of the world and used the fact of racial animosity and inequality in the U.S. as evidence of the failure of U.S. democracy closer to home. In this environment, the civil rights movement of the 1950s and 1960s utilized the television and photojournalistic accounts of racial violence to help build a personal connection between, on the one hand, southern blacks who were using nonviolence to resist the barbaric attacks of local police and angry white mobs, and, on the other, the white community outside the South. Taylor Branch described the powerful images of police dogs and fire hoses let loose on children as "rac[ing] towards" living room television sets across the country.[74] These images struck home, literally.

One of the main vehicles for communicating these images was the media generally, and presidents Truman, Eisenhower, and Johnson all expressed their concerns that images of racial unrest were being beamed throughout the world, giving public relations ammunition to the Soviet Union. Conscious of the role that the media played in swaying national and even international attention, the leaders of the civil rights movement intentionally crafted their strategies to attract media attention to their struggle. Often the best advertisements for the movement's position were the harsh and merciless actions carried out by the opponents of racial justice.

In the early 1960s, the civil rights movement plotted out specific, local, place-based strategies. One of their prime targets in those years was the city of

Birmingham, Alabama, where racial unrest had led to violence and racially motivated bombings of the homes of African American leaders and their churches. Groups in Birmingham, like the Alabama Christian Movement for Civil Rights, led by Fred Shuttlesworth, organized boycotts of local businesses that refused to serve African Americans. Knowing that his presence would bring national media attention, Shuttlesworth invited Martin Luther King, Jr., to help lead a concerted effort organized around Easter season, 1963.[75] Not only had Birmingham become identified as a center of racial violence, but the head of public safety for the city, "Bull" Connor, was known as a hot-tempered and extreme racist prone to reacting aggressively when confronted. Thus Birmingham seemed a good place to launch an effort to highlight and expose the evils of segregation and create a situation that would bring national attention to the cause of civil rights.

According to the Reverend Wyatt T. Walker, executive director of the Southern Christian Leadership Council, the feeling was that confrontational tactics would elicit a violent response from the local authorities and that the media would be drawn to the story and publicize it. That publicity would ultimately elicit national attention. Walker explained his media philosophy as follows: "My theory was that if we mounted a strong nonviolent movement, the opposition surely would do something to attract the media, and in turn induce national sympathy and attention to the everyday segregated circumstance of a black person living in the Deep South."[76] During a church rally, one of the movement leaders, Ralph Abernathy, declared: "The eyes of the world are on Birmingham tonight."[77]

Because Connor had already arrested many adults who had defied his orders regarding the political actions taken by the movement, the leaders needed a new group of activists to continue their efforts. Moreover, King lamented that the movement was losing the press, and needed to do something bold to gain the attention of the national media once again.[78] In response, civil rights leaders organized a march of schoolchildren in support of those movement activists who had already been arrested and jailed. To counteract the overwhelming turnout of students, Connor had his forces train police dogs and fire hoses on the children. Still photographs captured the melee that ensued: one of a police dog, teeth bared, being restrained by a police officer to keep him from lunging at a teenage boy; another of fire hoses being trained on three children. After these images emerged in the media, more journalists and television news cameras descended on Birmingham and the televised images of confrontations were circulated throughout the nation and even the world.[79]

Taylor Branch described a growing awareness, both inside the movement and from national leaders, that the exposure of the segregationists' harsh tactics in response to the actions of the protesters was both galvanizing local support for the movement as well as breaking the resolve of the Kennedy administration to refrain from intervening in the Birmingham crisis. As Branch explained: "Hard upon this surge of internal strength radiated the national news that a thousand Negro children had marched to jail in two days, and before the far-flung American public could begin to absorb such a troubling novelty, violence, the universal messenger, was racing toward their living rooms with pictures of water hoses and dogs loosed on children."[80] Walker described this media attention to the violent response of the public authorities as a "coup," and one that the movement could not have accomplished on its own with limited resources.[81]

Realizing the power of television, and extending the civil rights struggle to the airwaves, Medgar Evers pressured the local television station in Jackson, Mississippi, to allow him television time to respond to the editorial commentary the station aired that was critical of the civil rights movement. After meeting stiff resistance, a local station finally afforded Evers some airtime. With the opportunity, he gave his speech entitled "I Speak as a Native Mississippian." Speaking about a "Negro plantation worker in the Delta," Evers talked about the images that worker saw on the television and juxtaposed those against his experiences on the streets of Birmingham: "He can see on the 6:00 o'clock news screen the picture of a 3:00 o'clock bite by a police dog." He continued: "He knows that Willie Mays, a Birmingham Negro, is the highest paid baseball player in the nation. He knows that Leontyne Price, a native of Laurel, Mississippi, is a star with the Metropolitan Opera in New York. He knows about the new free nations in Africa and knows that a Congo native can be a locomotive engineer, but in Jackson he cannot even drive a garbage truck."[82]

When an agreement had been reached to try to bring the violent confrontations in Birmingham to a halt, a month after Evers's televised speech, President Kennedy gave a nationally televised address to announce the measures the federal government would take to reduce the likelihood that renewed unrest might jeopardize the settlement of the dispute. He also stated, with few details, that he planned on referring civil rights legislation to Congress soon. Tragically, however, just hours after Kennedy's address, the depth of the crisis in the South was made clear by the assassination of Medgar Evers outside his home in Jackson, Mississippi.[83]

Following Kennedy's assassination just five months after Evers's, President Lyndon Johnson shared the civil rights movement's appreciation for the power of the media and images in swaying popular opinion. A telephone exchange, recorded in January 1965, between President Johnson and Martin Luther King, Jr., illuminates three things about Johnson's approach to and understanding of the civil rights struggle: he believed in equality, he appreciated the role of the media in bringing about social change, and he knew that advances in civil rights were critical to his administration's foreign policy. The two men were discussing the importance of passage of voting rights legislation. In addition to Johnson's desire to permit the U.S. Postal Service to register voters, they discussed the need to get the message out about the barriers African Americans were facing to voting in the South.

After some initial discussions, King references the importance of placing an African American on the president's cabinet and he notes the impact on the public image of the U.S. on the international stage. "We feel that this would really be a great step forward for the nation, for the Negro, for our international image, and do so much to give many people a lift who need a lift now," King says, "and I'm sure it could give a new sense of dignity and self-respect to millions of Negroes who—millions of Negro youth who feel that they don't have anything to look forward to in life."

Later in the exchange, Johnson stresses the voting rights should apply to everyone—that this was not a question of special treatment for one class of citizens. "I think it's very important that we not say that we're doing this and we're not doing [this] just because it's Negroes or whites," he would say, "but we take the position that every person born in this country, when they reach a certain age, that he have a right to vote, just like he has a right to fight [*chuckles*], and that we just extend it whether it's a Negro, or whether it's a Mexican, or who it is." He continued, "I think that we don't want special privilege for anybody. We want equality for all, and we can stand on that principle."

Johnson also showed he had a clear ability to identify stories that could help illuminate the problems African Americans were facing and how King and his supporters had to get those stories out into the media and communities across the country. Johnson would say that King "can contribute a great deal by getting your leaders and you, yourself, taking very *simple* examples of discrimination [through voter tests] where a man's got to memorize [Henry Wadsworth] Longfellow, or whether he's got to quote the first ten amendments, or he's got to tell you what Amendment 15 and 16 and 17 is, and then ask them if they know and show what happens, and some people don't have to do that, but when a Negro comes in, he's got to do it."

Johnson then explicitly references the propaganda tactics of the Nazis when considering how to promote the cause of the civil rights movement: "If we can just repeat and repeat and repeat—I don't want to follow [Adolf] Hitler, but he had an idea—." For Johnson, that idea was "if you just take a simple thing and repeat it often enough, even if it wasn't true, why, people'd accept it." For Johnson, though, there was nothing dishonest in the cause of civil rights. He continued: "Well, now, this is true, and if you can find the worst condition that you run into in Alabama, Mississippi, or Louisiana, or South Carolina . . . [of someone] being denied the right to cast a vote, and if you just take that one illustration and get it on radio, and get it on television, and get it on . . . in the pulpits, and get it in the meetings, get it every place you can, pretty soon the fellow that didn't do anything but follow—drive a tractor, he'll say, 'Well, that's not right, that's not fair.'"[84]

Months later, the attack on marchers on the Edmund Pettus Bridge outside Selma, Alabama, may have been the final turning point in galvanizing political support behind the cause of civil rights. The televised images of those attacks were clearly at the center of this ultimate shift in the national mood. As Aniko Bodroghkozy describes in her book *Equal Time: Television and the Civil Rights Movement*, 48 million Americans were tuned in to ABC's airing of a new, star-studded broadcast: *Judgment at Nuremburg*.[85] The cinematic depiction of the trial of Nazis for war crimes perpetrated during the Holocaust, which also explored the complicity of German citizens in the genocide, had just started when the network broke into the telecast with a "breaking news" report from Selma. That report was a live broadcast of the attack on the nonviolent marchers. The juxtaposition of the racial violence in Alabama and the trial of Nazis for carrying out the Holocaust could not have been lost on millions of U.S. viewers sickened by the attacks. Soon after the watershed event in Selma, and the broadcasts of that day's events, President Johnson himself took to the airways to make an impassioned televised speech to the nation in support of voting rights.[86] Months later, Congress passed, and Johnson signed, the Voting Rights Act.[87]

The civil rights movement gained notoriety and support from the presence of the national media, particularly television, when the press covered the movement's activities. There are several other ways in which the relationship between the movement and television was both deep and symbiotic. First, despite the blatant acts of violence carried out against movement participants, often caught on film, many within the movement felt that the presence of the media helped temper this violence. As Ruby Hurley, an early leader in the movement, explained, the 1950s—"when there were no TV cameras with me to give me protection"—were far worse than the 1960s in terms of racial violence because of the presence of the media in that later period.[88] Second, at a time when network television stations were being criticized for the lack of substance in their programming, and televised news was still in its infancy, coverage of the civil rights movement gave the industry a subject with moral weight as well as one that generated powerful images, lending credibility to the medium.[89] Third, television itself generated tools for the movement to recruit and train new members. As part of the desire of television networks to bring legitimacy and gravitas to their work product, many networks produced documentary films on important subjects. One such documentary, a film produced in 1960 entitled *Sit In*, highlighted the activities of movement activists. Movement organizers later used the film as a training video, and it provided inspiration to activists.[90] Finally, a national television market raised the profile of activities of the movement in protesting oppression of the African American community. Indeed, what had been essentially local interactions—the abuse of blacks by whites—turned into a national issue as nationally televised acts were available, in the words of Russell Baker, when talking about the horrors of the war in Vietnam, "in the living room corner . . . at the touch of a switch."[91]

Thus the connections between the civil rights movement and the latest available communications technology—the television—are deep and varied. As the television rose in prominence and penetrated the market, the civil rights movement gained strength and support. With TV connecting the personal to the political by projecting images into people's homes, consciousness, and consciences, it became difficult for many Americans to look away and ignore the problems. Leaders of the movement used television and other forms of media to great effect, orchestrating tactics and strategies for maximum media attention, knowing that such attention would help expose the roots of racism and the horrors of Jim Crow.

Despite this ability to project images from the front lines of the civil rights movement's activities into virtually every living room in the U.S., the movement as a whole, as evidenced by groups like the NAACP and the Freedom Riders, was formed into different groups, and those groups typically organized into local chapters that were connected throughout the nation in strong networks. That is, they were translocal, often integrated to a certain extent by race, and often cross-class as well. As we have seen in previous chapters, these connections—between technological innovation, organizational structure, and social change—are common. The question that remains, and which I will explore further in coming chapters, is whether they are inextricable.

Technological and social innovations contribute to social movements in critical ways, and appear to have helped shape those movements. Advances in the ability to communicate—whether through the written or spoken word or an increased ability to travel—all had a profound impact on the social movements that arose in the wake of those advances, creating the social innovation moments I have identified so far. Furthermore, as advances in the means of communication were made, they shaped the movements that utilized and deployed them. Indeed, advances in the ability to communicate themselves spurred advances in civic engagement and social movements, helping to bring about social change. What is more, social movements in such moments often organize into social structures that mimic the means of communication they utilize.

Simply put, groups often end up looking like the means of communication they deploy. Social movements that arose in the early nineteenth century often took forms that replicated the other great social innovations of this period: i.e., the federal government, and, more specifically, the post office itself, outposts of which could be found in practically every town and village, regardless of the wealth of the inhabitants. As the media theorist Marshall McLuhan said: "Each form of transport [of goods and information] not only carries, but translates and transforms the sender, the receiver and the message." Indeed, as McLuhan asserted, and as the history recounted in this chapter would seem to expose and subsequent chapters will only reinforce: "The use of any kind of medium or extension of man alters the patterns of independence among people, as it alters the ratios among our senses."[92] Thus the means of communication shaped the very form of the organization that social movements would adopt. This, in turn, would ultimately have a profound effect, on many levels, on the nature of the change these networks would pursue and accomplish.

But it is not just technology that brings about social change in social innovation moments. The social movements that arose in the wake of these advances in communications technology tended to do things other than merely harness communications technology. They came together in countless town squares, union halls, community centers, churches, and synagogues and created trust in these face-to-face settings. They lowered the social distance between different communities by enabling them to see the common humanity shared across distance. By virtue of the most modern means of communication available to them, they also mostly utilized a networked, translocal structure and embraced broad-based and inclusive messages. To what extent are all these phenomena connected? The next two chapters explore this question.

2

NETWORK

Just as abolitionist leaders were beginning to harness steam power to spread their message, they also felt it was time to organize a national entity that could bring together some of the small and far-flung antislavery groups spread throughout the North. Although there were some state-based umbrella groups of organizations, most abolitionist societies were local and had little influence outside of their immediate community.[1] Abolitionists gathered in late 1833 in Philadelphia to create the American Anti-Slavery Society; they published a "Declaration of Sentiments" that echoed the language of the Declaration of Independence and found the injustices of King George III paled in comparison to those of slavery.[2] Following this convention, representatives from different state-based entities began to organize in earnest, forming local chapters through the northern states, with much of this effort taking place in New England, Pennsylvania, Ohio, and New York. One report of the extent of the reach of the American Anti-Slavery Society and its state subdivisions identified thirteen hundred local chapters spread throughout the North, with over a hundred thousand members.[3] The effort to organize these local chapters was carried out through "agents": individuals who would go town to town and engage in face-to-face conversations to bring supporters into the fold, pressing the abolitionist cause, encouraging local groups to form, and providing advice on adopting chapter constitutions. Inspired by the approach of Evangelical preachers during the Second Great Awakening,[4] these individuals could gauge support or opposition, form personal relationships, encourage local engagement, develop grassroots leadership from within local communities, and build trust through in-person interactions throughout the countryside.[5] As a leading abolitionist William Lloyd Garrison asserted, this local organizing was critical to the cause. He proclaimed his "*deep conviction that without the organization of abolitionists into* [local] *societies* THE CAUSE WILL BE LOST."[6] By meeting in person and face-to-face with those who might be interested in supporting abolitionism, he asserted that he was "able to disarm whole communities of their antipathies and rally them around the [abolitionist] standard . . . who else might remain indifferent or hostile to our cause."[7]

The different antislavery societies were organized in a network structure, with local chapters and statewide umbrella groups that linked up with national entities. One group of female abolitionists, the Boston Female Anti-Slavery Society, systematized the functioning of its network, with local canvassers going door-to-door to gather signatures for its petitions. These canvassers would then communicate with a local ward captain, who then communicated with a local precinct leader, and so on up the hierarchy of the network. The leaders collected the petitions and delivered them to a local, state, or national legislator's office.[8]

But the formation of these abolitionist groups into local chapters that were embedded in larger networks tells just part of the story. The groups that formed were also cross-class, as Edward Magdol's study of abolitionist societies showed. Factory workers, artisans, skilled laborers, and shopkeepers all joined these societies, signed petitions, circulated abolitionist tracts, and fought for the end of slavery.[9] Although African Americans and women often spoke at abolitionist rallies, many of the groups themselves might have been further subdivided along racial and gender lines, with women's auxiliaries and groups made up of free blacks emerging as well. These groups and their leaders worked collaboratively to further the abolitionist cause.[10] As the historian Richard Newman described these collective efforts, "Grassroots events brought together the kaleidoscope of Americans that new-style agitators hoped to energize—black and white, male and female, young and old."[11] Since the goal was to affect public sentiment, the field organizers began to realize that attracting a wide cross-section of members to the local societies had a real value in the overall advocacy and effectiveness of the movement.[12]

From the Sons of Liberty to the civil rights movement, different social movements have utilized different means of communications to form, identify individuals who shared common beliefs, bind people together, and animate collective action. They often did this in networks, and networks of a particular kind. In this chapter I explore the role of networks in social change. I focus on how trust is a critical component of that change and the ways in which a network of a particular kind—the translocal network—is an effective means of organizing movements in order to produce the trust and cooperation any movement needs for sustained effectiveness. As the name implies, the translocal network is one that may span a wide geographic area and can harness the power of a large group of committed individuals but is also made up of smaller cells where face-to-face communication between individual members can occur. The last chapter described how different media have promoted the growth and spread of these translocal networks, which, in many ways, mimicked the means of communication those networks used to communicate. We now turn to why such translocal networks are critical to advancing social change, how social movements face collective action problems, and how such problems are addressed through trust and trustworthy behavior. Such trust is present in the social relationships one forms with others, what is often referred to as one's social capital. This social capital can be harnessed for solving collective action problems when it is activated and carried out through networks of trust. What I hope to show is that translocal organizations have proven successful in harnessing this networked trust because such trust—the glue of social movements—is more likely to arise in translocal settings.

What follows is a description of the role of trust in solving collective action problems and the ways in which translocal networks harness this trust to advance broad social change. I will use examples of the social movements that have emerged in social innovation moments to show how they have generally tended to organize themselves into translocal networks, at least until a means of communication emerged that allowed organizers to form different types of organizations. What this chapter and the next explore is what network theory, social capital theory, and social movement theory all have to say about the reasons translocal networks may have proven successful, over time, in bringing about social change. The drift away from this translocal form of organization and this style of organizing, spurred on by a new technology, is the subject of chapter 4. What all of these chapters have in common is their exploration of the role of organizing and organizational structure in promoting lasting social change. At the core of this analysis is the notion of trust.

Social Movements as Collective Action Problems

When social movements form, they often come together to address what are known as "collective action problems": knotty issues that require cooperation to resolve them but where cooperation might, itself, be difficult to achieve.[13] Breaking free from colonial rule, abolishing Jim Crow laws, overcoming sexism to grant women the right to vote: these are collective action problems, problems that require cooperative, collective solutions. Individuals must work together to build support for political and social change, sometimes through mass resistance, sometimes through the ballot box or lobbying, sometimes through a combination of these and other strategies. Group cohesion; individual sacrifice; and sustained, coordinated action are all essential elements of collective action solutions and successful campaigns for lasting social change. The paradox of such solutions and campaigns is that it is sometimes in an individual's short-term interest to let others expend the effort to support the cause and hope the campaign will still succeed through the work of others.

Theorists for millennia have considered how to address collective action problems where cooperation is essential, but self-interested behavior can undermine such cooperation. Aristotle described this phenomenon as follows: "For that which is common to the greatest number has the least care bestowed upon it. Everyone thinks chiefly of his own, hardly at all of common interest; and only when he is himself concerned as an individual. For besides other considerations, everybody is more inclined to neglect the duty which he expects another to fulfill."[14]

A widely accepted view in the social sciences in the mid-twentieth century was that individuals acted rationally to further their own ends and that groups of individuals acted rationally, together, to further their interests collectively.[15] Whatever was in the best interest of the group as a whole is what the group would pursue. This vision of the workings of the public policy arena posits that different groups with different interests do political battle with each other based on these interests and the appropriate outcome flows from this process, a result of the strength of the various groups in relation to each other. This outcome typically reflects both compromises made among the groups but also serves as an indicator of the groups with the strongest general support. In his 1965 book, *The Logic of Collective Action*, the economist Mancur Olson challenged this view. Groups are not always a simple reflection of their members' rational self-interest that is multiplied by the number of participants. He would argue that groups that might otherwise enjoy wide support do not always achieve the outcomes they seek in the policymaking realm precisely because the larger the group, the more likely it is that group solidarity will break down and the "rational" thing the member of a group might do is not participate in the group's activities, believing others in the group will do the hard work necessary to achieve the group's ends.[16] Olson described this as the "free rider" problem, and there is hardly anyone who has ever embarked on group work that has not seen this phenomenon occur. In social movements, if enough people check out and the rest of the group is aware of such behavior, it can create a vicious cycle: those still engaged in the group's activities will become demoralized and will not want to be the ones carrying the dead weight of those who are doing nothing. In the end, if the free rider problem grows large enough or pervasive enough in the organization that even the individuals most committed to the cause abandon the work because of the high rate of defection of others, the effort fails.

Several years after Olson first published his *Logic*, population scientist Garrett Hardin, in his classic essay for *Science* magazine,[17] came up with a metaphor that describes a similar phenomenon involving collective action, what he called the "tragedy of the commons." What this thought exercise reveals is that in a world of limited resources, unless individuals are restricted in their behavior in some way, either by norms or rules that encourage cooperation, they will take advantage of a situation where the good to come to them individually may harm greater society. He used as an example a public pasture, where local herders would allow their cattle to graze. A degree of cooperation among the parties is necessary to maintain, preserve, and sustain the asset for productive, long-term use. A willingness to break from the crowd and use more than one's fair share of the resource can undermine this careful balance in the community. Once that occurs, there can be a rush to add cattle to one's herd for fear the resource will be depleted by other, noncooperative herders. As more individuals engage in the noncooperative behavior, the stronger the undertow to join them. The tragedy of the commons is a flip side of the free rider problem in some ways. The individual does not step aside and let others do the work; rather, he or she takes advantage of the goodwill and effort of others to try to get ahead at the cost of the long-term success of the venture. For both phenomena, where collective solutions and cooperation are necessary for a successful, sustainable, long-term, beneficial outcome for the group, the rational thing for the individual to do is to conduct a cost-benefit analysis to assess the relative value of cooperation and the effort necessary to cooperate in light of what he or she will receive in return. But short-term benefits can sometimes impede the prospects for long-term gain that comes with sustained cooperation. More important, one's lack of faith in the efficacy of collective solutions for obtaining long-term benefits may lead to defection, a withdrawal from the collective effort, creating, in a way, a self-fulfilling prophecy.

Obviously, though, not all cooperative ventures are doomed to failure, even in the types of settings Hardin imagined. Indeed, the Nobel Prize-winning economist Elinor Ostrom studied the management of common pool resources in communities that seemed to handle them well, basically exploring Hardin's commons metaphor in real-world settings. She concluded that smaller communities, ones that met certain "design criteria" as she called them, were better able to manage common pool resources than others. She found that smaller communities fared better in terms of those issues where there was a high degree of participation and engagement of all community members in resource allocation decisions, rule setting, and monitoring of participant behavior.[18]

Similarly, even Olson, who helped inspire a cost-benefit view of collective action, seemed to think that some organizations are capable of overcoming the free rider problem. In the large organizations that are generally necessary to bring about broad social change, monitoring of participant behavior is more difficult, and it gets even more difficult as group size grows.[19] At the same time, Olson posited that large groups that form as networks of smaller groups are in a better position to overcome this problem. The dynamics in these small groups are such that, when embedded in a larger network in what he called a "federated" fashion, they are able to have the impact that larger institutions can have, while still being able to ferret out the free rider problem at the local level. Social pressure to exact compliance is best applied in small group settings, and when those smaller groups are connected to a larger network, they are able to leverage the power of a larger group, while maintaining the ability to monitor individual participant behavior at the local level.[20] Ostrom agreed. Her design principles can still work well in larger institutions where there were smaller communities connected in what she called "nested enterprises."[21] For Ostrom, these were like Olson's federated groups: collections of smaller institutions "nested" in larger networks that combined to form local, state, regional, and national systems.[22]

For Olson and Ostrom, these nested or federated groups function better and help monitor the free rider problem more effectively for a simple reason: individuals in such settings are more trusting and trustworthy. When individuals are able to band together in small groups because they trust one another, they will not take advantage of others because they want to behave in a trustworthy fashion. But why is that? And does available research help us understand the situations in which trust and trustworthiness are more likely to arise? Hardin's commons metaphor is an example of the application of a game theoretical approach that might help us understand social dilemmas, how we behave within them, and how certain settings may be more conducive to trust and trustworthiness. Economists and philosophers have used game theory as one avenue to assess the most rational or logical choices individuals are likely to make to maximize their own position in certain settings. Game theory is sometimes used to identify the conditions that are more likely to generate trust, and the findings of some of these studies can be used to identify the role organizational settings and structures can play in generating cooperation.

Some common game theory applications include the prisoner's dilemma and the dictator game. In the problem of the prisoner's dilemma, we are asked to engage in a thought experiment to consider a situation where two prisoners have been arrested for the same crime and are placed in separate interview rooms by the police. The best approach for both prisoners is to say nothing to the authorities. If they remain silent, neither will face stiff jail time. If one "defects," i.e., rats out the other, he or she might go free, while the other faces conviction for the crime. The question becomes one of trust: Will the two trust each other enough to remain quiet and not turn the other prisoner in?[23] Similarly, one of the manifestations of the dictator game involves distribution of a sum of money, with one party able to give money to the other with the hope that a portion will be returned to him or her. As in the prisoner's dilemma, the dictator game revolves around how much the participating parties will trust one another.

Game theory experiments provide insights into the role of trust in cooperative actions, which yield further insights into the workings of social change and the movements that seek to bring it about.[24] The first of these insights is that participants will cooperate more when they believe they will be paired with the same partners on multiple occasions: what is referred to as "repeat play."[25] In a series of experiments, Robert Axelrod invited several hundred individuals to participate in a prisoner's dilemma setting ("PD" for short) to see how they might act within it in an effort to determine those strategies that generate the best outcomes in the game.[26] Axelrod used an "iterated" version of the PD game, where parties engage (and know they will engage) in multiple transactions together over time. Axelrod found that the best outcomes occurred through what he called a "tit-for-tat" approach, where each player mimics the prior player's move, cooperating in the face of cooperation or defecting as a form of punishment to retaliate for prior defection. The initial action designed to produce the best results and the most cooperation is the "trusting first move," which encourages and induces one's counterpart to reciprocate in a cooperative fashion, leading to a virtuous cycle of cooperation.[27] Cooperation breeds cooperation and defection—choosing the selfish outcome—breeds defection.[28] We cooperate when others cooperate with us; we choose not to cooperate when others are doing the same.[29] This should come as no surprise. But the notion of the trusting first move presupposes that there will be subsequent moves between the players. In repeat play settings, the incentives are aligned to encourage cooperative behavior and punish noncooperative behavior: when newly formed pairs begin to cooperate, when noncooperative behavior is punished, and when forgiveness is shown the offender who exhibits a willingness to stay on the straight and narrow in the future. Such behavior tends to foster cooperative behavior among repeat players who are paired over a series of games.[30] With repeat games, a cooperative "first move" by a partner in a newly paired team tends to trigger a virtuous cycle of mirrored behavior despite the risk of exploitation.[31] If tit-for-tat tactics are common, cooperative first moves result in positive outcomes in the long run in repeat games, as cooperation and trust begets more of the same.[32] Central to these phenomena is the notion of repeat play. We cannot generate the benefits of a first trusting move or tit-for-tat tactics if there will not be extended interactions between potential cooperators.

A second key insight about cooperation and trust that game theory research reveals is that greater social distance increases untrustworthy behavior and makes cooperation and trust less likely. Social distance between individuals lowers trust and increases the chance that individuals will try to take advantage of each other.[33] While the thought may be unpleasant, we trust those we perceive as similar to ourselves, those we classify as being members of the same groups.[34] We even tend to trust those who look more like us than those who do not.[35] Social distance arises based on distinctions in race, gender, nationality, and language, among a wide range of other differences. From cross-country and cross-community analyses we learn that nations and communities with more heterogeneous populations have lower generalized trust, and as that heterogeneity increases, trust decreases.[36] Another measure of social distance is income inequality, and societies with greater income inequality also have lower trust.[37] A perfect example of this phenomenon is the United States: generalized trust in the U.S. has declined in recent years as income inequality has increased.[38]

A third insight about trust to emerge from the research is that communications and expressions directed toward encouraging cooperative behavior lead to greater trustworthiness. In one of the earliest attempts at the classic prisoner's dilemma experiment, when subjects exchanged notes promising cooperation in advance of the exercise, they tended to cooperate at a higher rate than those participants who did not.[39] Research also shows that mere communication between the parties, even without a promise to cooperate, can increase empathy between parties, decrease social distance, and lead to greater trust and cooperation regardless of such distance.[40] What is more, even unilateral communications involving trustworthiness can generate more trustworthy behavior. In one study, when subjects affirmed that they knew a test was being administered under a school's honor code they tended to cheat less, even where such an honor code did not exist.[41]

What game theory can tell us is that there are certain conditions that make cooperation and trust more likely. When individuals have opportunities to engage in repeated interactions, and believe they are going to interact in the future, they are more likely to cooperate since cooperation is seen as in their best, long-term interests. The lower the social distance between individuals (which is often an effect of close interactions and communications), the more likely it is that they will trust each other and engage in cooperative behavior together. Finally, when individuals communicate with one another, whether they are expressing a willingness to cooperate or simply getting to know one another (which can lower social distance), they tend to cooperate more. If what game theory tells us about the conditions in which cooperation is more likely to occur is true, it does not take much of a leap to conclude organizations that have a local, face-to-face component are more likely to generate trust, trustworthiness, and cooperation because these are the settings in which the trust-generating features described earlier are most likely present: repeated, positive interactions; decreased social distance; and communication. This ability to tap into these feelings of trust and a community of cooperation they can foster is often considered one's social capital, and this form of capital, it turns out, may be essential to social movement success.

Social Capital Theory, Cooperation, Networks, and Trust

Social capital is said to be found in the "social networks and the . . . norms of reciprocity and trustworthiness" associated with such networks.[42] These networks and norms lead to cooperation. They generate feelings of mutual obligation toward other members of a group, encourage information sharing, and convey a code of conduct that carries sanctions for a group member's violation.[43] The term "social capital" was first used by a West Virginia educator, L.J. Hanifan, in a 1916 essay for the American Academy of Political and Social Science. When an individual comes in contact with a neighbor, and then with other neighbors, Hanifan wrote, "there will be an accumulation of social capital, which may immediately satisfy his social needs and which may bear a social potentiality sufficient to the substantial improvement of living conditions in the whole community."[44] In his landmark work, *Bowling Alone: The Collapse and Revival of American Community*, Harvard's Robert Putnam elevated the term "social capital" to the national discourse. Putnam argues that "life is easier" when social capital is high in a community.[45] He explains: "Networks of civic engagement foster sturdy norms of generalized reciprocity and encourage the emergence of social trust."[46] These networks help communities coordinate and communicate, spread information about reputations for trustworthiness in a community, and help communities solve problems. For Putnam, there is less cheating in such communities because "economic and political negotiation" takes place in "dense networks of social interaction."[47]

Social capital theorists identify two types of social capital. "Bonding" social capital forms in closely knit groups. This type of social capital facilitates information sharing, mutual support, and collective action. The sociologist Xavier de Souza Briggs has described this type of social capital as "social support," which "helps one 'get by' or cope."[48] This version of social capital "might include being able to get a ride, confide in someone, or obtain a small cash loan in an emergency."[49] By contrast, "bridging" social capital facilitates connections across different close-knit networks.[50] For de Souza Briggs, this type of social capital is "social leverage," which "helps one 'get ahead' or change one's opportunity set through access to job information, say, or a recommendation for a scholarship or loan."[51] Bridging social capital allows individuals and networks to move beyond close-knit bonds and share information and ideas across otherwise insular networks.[52] Bonding social capital within a community can provide critical support to members to find local jobs, help with day-to-day needs, and offer financial support. Bridging social capital helps form linkages across groups, providing access to new networks, expanding job prospects, and increasing the reach and strength of bonded groups.[53]

Social capital can arise in both forms in the same setting, and individual members of a group where bonding social capital is present can support each other by bringing their own connections, their "bridges," to bear in order to support members within the group. When a member of a community group leverages her connections at her place of employment to help a fellow member of the group find a job, both bonding and bridging social capital is coming into play. One study of a U.S.-based microenterprise program showed how these different types of social capital can impact cooperation. The program was designed to build on and leverage both bonding and bridging social capital to support program participants. Borrowers formed mutually beneficial relationships with the other borrowers in the group. The bonding between the members was manifest when they shared information about promising practices, and these supportive relationships helped build members' self-esteem. In turn, the members formed relationships with other organizations, outside lenders, and government entities. These bridging relationships also increased economic opportunities for the members of the group.[54] The findings about these benefits confirm the notion that it is through bridging social capital, more than bonding social capital, that we do "get ahead." The sociologist Mark Granovetter dubbed this phenomenon the "strength of weak ties": the fact that individuals often improve their lot more through bridging social capital than bonding social capital, gaining more benefits from their looser connections across social networks than they gain from their more immediate circle of contacts.[55]

Of course, social capital can have its "dark side." Not all civic associations generate positive interactions and broader benefits for society. The Ku Klux Klan and organized crime organizations reflect some aspects of social capital at work. These networks are marked by social interactions and trust working toward a common goal.[56] Like other forms of social organizations, these networks can serve to bring mutual aid to their members and may even serve as a source of job and business opportunities, if we can call them that. Despite the social and economic benefits members of such groups may reap from membership, which may look identical to the benefits more constructive groups may yield for their members, in these manifestations, social capital does not generate benefits for broader society. In fact, too much bonding social capital can make a community insular and distrusting of nonmembers. And some research suggests that communities and nations that are less diverse tend to have more social capital. It may come as no surprise that trust is generally higher in more homogeneous communities where social capital is high.[57] Similarly, research suggests that in more heterogeneous communities, social capital can be harder to develop.[58] At the same time, communities can create social capital even in heterogeneous communities when they are less segregated and there is mixing between groups otherwise separated by social distance.[59]

Putnam's views on the value of social capital, and his arguments that it has been in decline over the last few decades (the idea that we are now "bowling alone"), are not without their critics, however. Some cheer the decline in participation in self-selecting groups that might have kept out members of marginalized communities.[60] Others claim that civic engagement is not on the decline, but rather the nature of civic engagement has just shifted, from quaint bowling leagues to other, more modern forms of organizations and organizing.[61] What is difficult to argue with, however, as Theda Skocpol has shown, is that membership in traditional, translocal groups has declined considerably since the 1970s, a phenomenon I will explore in greater depth in chapter 4.[62] Even if we use Putnam's view of social capital, the decline in participation in traditional civic groups can have ramifications for democratic institutions and practices. The benefits of social capital extend beyond those that accrue to groups of neighbors engaged in civic associations themselves. Putnam's research exploring the history and strength of civic life and the effectiveness and responsiveness of government in Italy reveals some of the broader benefits of social capital. By studying civic participation in that country over time, he showed that communities with a history of robust civic activism and strong reserves of social capital had more effective governments than those with neither.[63] The loss of the spillover effects of the social capital such groups generate is what Skocpol calls "diminished democracy."[64]

How do these forces relate to trust and social movements and how does their absence diminish democracy? Olson raised concerns about the risks inherent in group behavior, such as the likelihood that, when given the chance, many humans will choose to step aside when it comes to collective undertakings in the hope that others will pick up the slack and do the hard work necessary to bring about an otherwise desired outcome. He called this the "logic of collective action." But when individuals do manage to come together in situations where they can communicate, plan for work they will undertake together into the future, and reduce social distance in the process, they can also trigger what the legal scholar Dan

Kahan called the "logic of reciprocity." When groups undertake collective action in the pursuit of solving collective action problems, through trust and reciprocal, trusting actions, individuals can spur further cooperative behavior.[65] Tocqueville described this phenomenon as follows: "One is occupied with the general interest at first by necessity and then by choice; what was calculation becomes instinct; and by dint of working for the good of one's fellow citizens, one finally picks up the habit and taste of serving them."[66]

One study of organizations that were mobilized to reduce drunk driving found that those groups in which the members of local chapters were given work to do on behalf of the organization, where they would come in contact with other members in face-to-face interactions, were ultimately more successful and effective in mobilizing those participants generally than those groups where such local activities were not a part of the group's efforts.[67] Similarly, as researchers who studied the activities of five different community mobilizations showed, individuals who engaged in face-to-face encounters with other members of the group and were involved in conducting collective research and planning for larger actions were more likely to continue their engagement with the group as opposed to those who only took part in large-group activities, the latter type of activity offering fewer opportunities for meaningful face-to-face interactions with other members.[68]

As Gordon Whitman, a long-time organizer with the national group Faith in Action, explains:

> It takes hard work and honesty to make social justice teams work. Small groups succeed when they create shared norms, build trust through frank conversations, adopt ambitious goals, create clear roles for people, and make important decisions together. With these ingredients, it's possible to organize large numbers of teams that can function effectively without depending on lots of paid staff. Within groups, people can follow a process of sharing stories, reflecting on text, and acting together that helps teams both get things done and be transformative for their members.[69]

So how do these forces relate to social movements? Trust is fostered in settings where social capital is cultivated, and from those settings social movements can arise. As the former Obama administration official Anne-Marie Slaughter writes: "Cooperation requires a basic level of trust. Trust, in turn, requires repeated human interaction, building a reservoir of social capital that supports the propensity of human beings who know and like one another to self-organize into groups and associations."[70] The idea of repeat play is at the heart of this civic engagement when individuals come together with the express purpose of collaborating to solve a problem they all share, and effective organizations are those where cooperation will be repeated and sustained. It is within such settings where we can diminish social distance, particularly in diverse settings, where individuals of different backgrounds are able to work together, communicate, find shared interests, and work toward common goals. Similarly, communication is at the heart of settings that foster the creation of social capital, and it is hard to imagine individuals coming together to work toward common goals without some form of robust communication between them. In other words, all the conditions that appear necessary to foster trust are present in small group settings where people can communicate on a face-to-face basis; when those small groups are linked to broader networks, they can cooperate at scale to address collective action problems.

For these reasons, then, the trust and cooperation that emerges from a robust and diverse network are both essential to organizing work, but forming such networks is not just a tactic for the organizer, it should also be a goal. As Leslie Crutchfield, who has studied the common features of several modern social movements, argues:

> Great movements have at their core strongly connected grassroots members. Leaders of movement organizations understand they need to invest in building member relationships—not just between the members and the organization or movement, but among the members themselves. They nurture intense, personal bonds that engender trust and mutual obligation. Building on those bonds, they then encourage activists to collectively take charge in their communities to advance the cause at the local level. The network also becomes an end in and of itself.[71]

In my own work as a lawyer for community-based organizations, I was a decent enough lawyer, and believe I was able to manage the technical aspects of the work I was doing on behalf of my clients with competence, but crucial to my work was the notion that the long-term viability of the groups they formed, the bonds of trust they established between each other within the group and to individuals in other groups, mattered more than whether we were working in an efficient manner. Cultivating the rank-and-file members of the group, building up their willingness to participate in activities beyond the one-off action, took time and effort. Night after night, gathered in building vestibules or my clients' apartments, my clients and I would meet to discuss strategies, tactics, and next steps in the organizing. The important thing for me was ensuring there was a functioning organization with organic leadership for sustained activities rather than one that either did my bidding, or pulled off flashy, one-off events. The group—the network—is what mattered; it was, as Crutchfield argues, "an end in itself."

In fact, I knew that the strength of the organization over time was the bonds between the members and not my own skills as an advocate. A landlord or business owner could wait out the group with which I was working if there was a sense that the group was weak. That weakness in the organization's ranks meant that, eventually, there would be no group there to oppose the landlord or employer. If the group's adversary learned of dissent or fissures in the ranks of the group, without labeling what we were facing a "free rider" problem, that adversary typically had an intuitive sense that such a problem was present and it would just be a matter of time before the group effort failed. The adversary might even use tactics to buy off wobbly and wavering members with promises of some economic benefits for disavowal of and defection from the group. Where social capital theory and the experience of many community organizers come together is around an appreciation for the fact that effective organizations require group cohesion that helps overcome the free rider problem. But such efforts are not easy. They require groups to tap into those affective elements of community organizing—the emotions that can motivate individuals to participate and steel them for a struggle—and organization that can channel those emotions into coordinated action.

What importance do trust and trustworthiness have on this coordination, group cohesion, interpersonal cooperation, and social movements? A lack of trust means people will not cooperate with others; a lack of trustworthiness means no one will cooperate with the person who is not trusted. No social movement can bring about change if it cannot get people to cooperate to pursue common goals. Recently, Putnam has generated what he calls a "lean and mean" definition of social capital; it is the "social networks and the associated norms of reciprocity and trustworthiness."[72] If I may hone the definition even more: social capital is networked trust.[73] Looking at the connection between social capital and collective action problems, Putnam posits further that it is these networks themselves that "facilitate coordination and communication, amplify reputations, and thus allow dilemmas of collective action to be resolved."[74] In other words, social capital creates a platform on which trust and cooperation can thrive precisely because of the phenomena discussed earlier: the settings in which social capital tends to flourish are settings that create opportunities for repeat interactions and communications, and these interactions and communications tend to lower social distance. Importantly, these phenomena are all present in the face-to-face interactions that happen on the local level. But it is when those face-to-face relationships are linked to together to other sites where similar interactions are occurring that social capital turns into a network for addressing collective action problems.[75] For social capital theory, trust and cooperation are embedded in networks that foster communications, interpersonal interactions, and decreased social distance—that it is, as I have described it, networked trust.

If that is the case, is there something we can learn from the field of network science that can offer more evidence of the benefits of translocal organizational structures for solving collective action problems, launching social movements, and driving social change?

The Science of Networks

Like social capital itself, a network consists of the relationships formed between objects, often called nodes.[76] Perhaps the most important characteristic of a network is the number of nodes in that network.[77] The distance between the nodes serves as a second key characteristic of a network, and the closer in distance one node is to another, the more likely a network will form. This is known in network theory as the concept of *propinquity*.[78] Another characteristic of the nodes that can influence the likelihood they might form a network is their similarity, often referred to as their *homophily*.[79] In many ways, in a network involving human beings, lower social distance between individuals translates into greater homophily. This similarity, or perceived similarity, can influence the formation of networks.[80] Another characteristic of a network is the aggregate number of nodes and that number can determine the strength of the network. A large network generates "network effects": what has come to be known as "Metcalfe's Law," for computer scientist Robert Metcalfe. Metcalfe posited that the benefits to be derived from a network do not simply increase incrementally with a numerical increase of the number of nodes in the network.[81] Metcalfe noted that the benefits to be derived from networks grow exponentially.[82] One fax machine is worthless. Two fax machines create a network. Three fax machines triple the faxing capacity of the network. Adding a tenth fax machine increases the power of the network in an exponential fashion.[83] Finally, identifying the *position* of each node in relation to each other is also critical for understanding the strength and reach of the network. Each node can be found at or near a center or the periphery of a network[84] and can have many connections in the network or just a few. A particular node can also serve as a hub connecting two different networks or fill gaps within a network.[85] Connections between networks arise, particularly in the social movement context, through "past histories of group relationships as well as dense, overlapping, interpersonal networks that sometimes link groups to one another."[86] Bringing two networks together can increase their collective strength, reach, and durability, and can generate increased network effects.[87]

A fascinating take on the science of networks comes from the theoretical physicist Geoffrey West, who has explored the structures of the hidden networks that he says drive the metabolism of virtually every living thing and many human organizations, such as cities and businesses.[88] For West, evolution over billions of years has led to remarkably similar structures that help a range of organisms—from cells and trees to rabbits and humans—function by regulating complex processes these organisms need to function. These same processes are found in human organizations as well, from cities large and small, Fortune 500 companies, and emerging start-ups. These processes are invariably carried out by networks, which are hierarchical and branched. These networks possess three common features: they are "space filling," have "invariant terminal units," and are structured to make the flow of whatever that network is circulating efficient.[89] In other words, the networks cover the organism or entity, the ends of the networks are similar, and the structure of the network (hierarchical, branched) is such that it generates the efficient functioning of the organism. The product of evolution over billions of years suggests, for West, that these phenomena reflect a sort of consensus around the best way in which to structure an organism, and such structures allow for organisms and organizations not only to survive, but to flourish. He explains that "whatever was at play" that led to the common functions of many organisms "had to be independent of the evolved design of any specific type of organism, because the same laws are manifested in mammals, birds, plants, fish, crustacea, cells, and so on." All of these organisms require "the close integration of numerous subunits—molecules, organelles, and cells—and these microscopic components need to be serviced in a relatively 'democratic' and efficient fashion" in order to function.[90] Another critical component of such well-functioning networks is that they are often made up of what West describes as "fractals": if you "cut a piece out of . . . a [hierarchical] network and appropriately scale it up, then the resulting network looks just like the original." In this way, "each level of the network essentially replicates a scaled version of the levels adjacent to it."[91] The hierarchical, branched networks that have a fractal quality to them throughout their systemic structure do sound a lot like the nested, federated organizations we have been discussing all along. These networks are branched and hierarchical and have smaller components, like Russian dolls, found throughout the network and as you go out to their farthest reaches.

Understanding networks in these ways helps put into context some of the discussion of translocal networks. Nodes that are closer to each other in terms of their characteristics and geography enjoy stronger bonds than those that are less similar and less closely related geographically. At the same time, networks as a whole are stronger when they are larger and build on the strength of closer dyads and triads embedded within them. Those close relationships bridge greater distances when individual nodes fill gaps and link nodes to extend the number of nodes connected to a network. They can even link other networks together to form larger, connected networks to generate more powerful network effects. In other words, larger networks—the ones that are needed to solve complex problems like the functioning of a city or to operate the metabolism of a mammal—require both close bonds and bridging connections.

These same phenomena are needed for human networks to solve collective action problems, and many of the more successful social movements over the course of U.S. history, particularly in social innovation moments, have been networked—even fractal—phenomena. In many ways, movements that have a face-to-face component to them at the local level become a means of channeling and communicating norms of cooperation, including the norms of trust and trustworthiness. But it is when they link up to create a broader network that they begin to develop true network effects. Putnam emphasizes the role that both norms *and* networks play in the functioning of social capital. Network theory shows how norms of reciprocity, trust, and cooperation can exist to help solve collective action problems, but these are networks built on *both* bonding and bridging social capital to solve collective action problems.[92] Close relationships foster trust; larger-scale networks constructed of many close relationships help groups address collective action problems, at least ones where the solution demands coordination beyond just a small, tight-knit group of people (as most true collective action problems require). As described earlier, in the 1830s the abolitionists knew they needed to form a national network made up of local chapters to increase the power of their movement. Fifty years later, the women's movement would do the same.

Following the Seneca Falls Convention in 1848, women's movement leaders mostly engaged in agitation and speaking at public events, holding follow-up conventions in other states without spending much time engaging in local organizing.[93] They suspended some of this work during the Civil War but later began again in earnest, especially as Congress debated the proposed constitutional amendments following the end of the conflict, most notably the Fifteenth Amendment. While many leaders of the women's movement emerged from the abolitionist movement and would have supported the right to vote for African Americans, some went so far as to oppose suffrage for blacks if women in general would not also enjoy the right to vote. Starting in the late 1860s, the two most prominent national organizations were the National Woman Suffrage Association (NWSA) and the American Woman Suffrage Association, with divisions between the organizations centered around suffrage for African Americans (the NWSA opposed the Fifteenth Amendment because it did not include women).[94] The groups each began to form local and state-based subdivisions, mostly in the East and Midwest.[95] Roughly twenty years later, the groups merged into the National American Woman Suffrage Association.[96]

In the 1880s, the first generation of leaders who had burst onto the scene with the Seneca Falls Convention gave way to a new group of advocates. With this new generation of leaders, the tactics of the movement changed from merely engaging in public speaking and agitation to conscious and intentional community organizing and the creation of grassroots networks formed into local cells that were connected to larger, umbrella organizations, including statewide groups and then national organizations. They directed advocacy and energy toward local organizing and the creation of formal organizations rather than the general consciousness raising that had predominated in the years before the Civil War.[97] This local organizing also took on a different focus, moving from advocacy for a national constitutional amendment to work closer to home, seeking suffrage at the state level in states where the populace might be more amenable to the concept.[98] Borrowing the abolitionists' tactic of using agents, women's organizations sent speakers and paid organizers into communities to spread interest in women's rights, not just for the goal of raising awareness about women's suffrage, but with the explicit goal of organizing local chapters to support the effort. Moreover, focusing on legislative or ballot referenda initiatives at the state level meant the organizing took place at that level as well.[99]

Ultimately, after victories at the state level in numerous states, the advocacy brought about passage of the Nineteenth Amendment in 1920. What impact and effect did the shift of tactics, from general consciousness raising to local organizing, have on this success? In her study comparing the success of women's suffrage efforts in the United States with efforts in Switzerland, where women did not win the right to vote until 1970, Lee Ann Banaszak attributes the difference in outcomes to the strength of the local organizing efforts of the U.S. suffragettes. Even though, on a per capita basis, the Swiss women's movement was larger (taking into account the smaller size of the population as a whole), the U.S. movement was much more effective, and Banaszak found that this success was mostly a product of the fact that U.S. women were organized, and not just in amorphous, national organizations like the Swiss, but rather were formed into national networks with local and state-based subdivisions, where local leaders could be cultivated, and yet national organizations could spread innovative tactics that emerged from those local leaders.[100]

The local organizing form among women was quite popular at the turn of the twentieth century, whether it was in women's suffrage organizations; temperance societies; or the pursuit of similar community improvement, cultural efforts, or social endeavors. In 1925, Alice Ames Winter, an activist in Minnesota and the president of the national General Federation of Women's Clubs (which still exists today), wrote a guide that women could use in organizing these clubs. In her book entitled *The Business of Being a Club Woman*, Winter described the value of not just local organizing, but of linking local organizing to a large purpose and the methods for combining local organizing to larger causes. For Winter, when we join a club, "something comes into our lives greater than any one can get alone." Club participation "blends companionship, friendship, the wisdom that comes from many minds rubbing against each other, the inspiration that springs from such contact of mind with mind, and the efficiency that results from the combined effort."[101] For Winter, it was this combination of local relationships with a broader, national, and even international network that strengthened women's influence and reach. An individual club extends one's "individual powers," she wrote, by "Friendship," "Wisdom," "Inspiration," and "Power." In addition, "The Federation" as she called it, "extends the powers of the individual club by" the following: "More Friends," "Added Wisdom," "Keener Inspiration," and "Power that grows by leaps and bounds."[102] Interactions between club members "begin with more intimate contact between one local group and another, and gradually grow until they have found that well articulated organization that stretches its filaments all the way from the individual woman and club up through city and county, district and state, to the national organization and even beyond the seas."[103] She described the components of the General Federation of Women's Clubs as follows: "The clubs are the rim, the states, the spokes, but the General Federation is the hub that holds them all together. Without it, they drop apart and the wheel of our joint progress stops running."[104]

What Winter described was the combination of bonding and bridging social capital at work, the harnessing of trust at the local level that generates broad network effects. In an earlier historical example, we can also see the two forms of social capital working together across a network to solve a collective action problem—that of rallying colonists to rise up against the British troops at the start of the American Revolution—in the example of Paul Revere. The historian David Hackett Fischer has described Revere as "a great joiner."[105] He was not only a veteran himself and a Freemason, he was a member of several secret societies and frequented many of the taverns in the Boston area that were hotbeds of revolutionary fervor. These activities brought him into contact with many leaders of different organizations and associations in the region on the eve of the hostilities.[106] Ultimately, his world-historical ride through the countryside, from town to town, was not directed at random farmers he might come across along the way. It was targeted at the other connected individuals Revere knew, as Fischer explains:

> Paul Revere and the other messengers did not spread the alarm merely by knocking on individual farmhouse doors. They also awakened the institutions of New England. The midnight riders went systematically about the task of engaging town leaders and military commanders of their region. They enlisted its churches and ministers, its physicians and

> lawyers, its family networks and voluntary associations. Paul Revere and his fellow Whigs of Massachusetts understood, more clearly than Americans of later generations, that political institutions are instruments of human will, and amplifiers of individual action.[107]

To describe human networks in this way—as combining elements of interpersonal bonding between individuals and then bridging across weak ties to form larger networks—aligns well with the discussion of social capital and its two forms: bonding and bridging social capital. These forms function the same way that networks operate because social capital is always embedded in networks. Those networks can manifest themselves as either the bonding or bridging variety. When networks grow, they can also exhibit characteristics of both varieties, as the Revere example shows. When connected networks are linked by individuals who draw different social circles together to make, for all intents and purposes, a single, functioning network, this can serve as an example of bonding and bridging social capital appearing together in the same network.[108] As in the Revere example, these types of social capital can be deployed to address collective action problems in efficient and effective ways.

Furthermore, it is these "hybrid" networks (those that exhibit both bonding and bridging social capital) that have fractal qualities and most resemble what Olson called federated organizations and Ostrom described as nested enterprises. It is in such organizations that coordinated, cooperative action is most likely to arise and in which a group can avoid the free rider problem because of the presence of trust-enhancing conditions described earlier. And it is for these reasons that networks with such relationships can be activated to address collective action problems.

It seems apparent that larger, networked groups are able to develop the network effects that come from scale. But what about non-networked groups that remain local, that do not attempt to reach beyond local issues? One would think that they do not suffer the same defects that larger organizations face in terms of the free rider problem, and likely form strong bonds of trust based on their face-to-face interactions; at the same time, they are not able to capture the network effects or the "strength of weak ties" that a nested, federated organization might yield. Comparative research on the success of different organizational or structural models of social movements from across the globe tends to show that those groups that stress only local issues and fail to adopt a translocal approach—i.e., those that are not made up of local chapters nested in a broader network—find it harder to move beyond the formation of strong local bonds and fail to have any impact on broader public policy. As the political scientist Joyce Gelb's research assessing the success of different women's groups showed, organizations in the

United Kingdom tended to form small, close-knit groups that engaged in extensive face-to-face interactions but were rarely connected together in a broader network. This made broader political mobilization difficult.[109] This sort of nonnetworked, nonfederated structure of some women's organizations meant that they could not easily mobilize for coordinated action. This, in turn, could lead to, in the words of political scientist Jo Freeman, "political impotency."[110] While there is certainly a value in creating local organizations that provide mutual aid and support, when the group wants to turn to coordinated, political advocacy, that sort of organization cannot sustain it. For Freeman, many of these groups would flounder were they "unwilling to change their structure when they change their tasks."[111]

How social movements are structured appears to have some effect on the relative success of those movements that emerge in social innovation moments. Groups that are organized in translocal, federated networks seem to have a proven track record of some success in bringing about social change, from the abolitionists of the nineteenth century to the civil rights movement of the twentieth, partly because such national networks tended to be organized in these translocal structures by necessity. There were no means of communication that facilitated one-to-many communications from leaders to individual members with ease. But there is one final component of what facilitates the functioning of effective networks for social change. Our engagement within social organizations and as parts of networks sends messages to others about our viability as cooperative partners. But messages also send messages. In the next chapter, I will explore how one essential element of social movements—the messages they promote—may, when building on the means of communications and the networks that build on those means, foster social movement success. Social capital theory and network theory can serve to explain some of the reasons for social movement success. The following discussion builds on these ideas and introduces social movement theory into the analysis to show how the messages that pulse throughout a network not only mobilize individuals, but also enable them to identify with a movement and foster the emergence of broad-based coalitions that are essential to effective and far-reaching social change. It is the message that makes up the third component of the social change matrix.

3

MESSAGE

In late 1943, the newly elected national commander of the American Legion, Warren Atherton, began his tenure by attempting to grapple with the problems faced by disabled service members returning from combat. The most urgent problem facing the Legion was for "right then, not later."[1] From Atherton's perspective, "hundreds were being released weekly without a cent in their pockets and no provision for support or care."[2] For many returning service members, their records of service had been destroyed in battle incidents. "Under then existing laws," Atherton explained, "the services were not paying dischargees whose records weren't in order. Service pay for some was a year and more in arrears."[3]

Atherton sent telegrams to local chapters of the Legion seeking examples of service members who had been having trouble accessing benefits on their return stateside. In twenty-four hours, he had more than fifteen hundred stories of disabled service members who had been waiting months for financial support from the government.[4] From the stories that were collected, the Legion produced a report citing these individual stories of veterans who had returned from the war and were having a hard time integrating back into the community, facing delays awaiting assistance and enduring severe economic hardship as a result. This skillful use of personalized stories helped draw support for the Legion's efforts because it made the hardship seem real.

The Legion's report served as political ammunition as its members and supporters marched on Capitol Hill to garner support for improving aid to veterans. Hundreds of newspapers supported the Legion's efforts with editorials and covered its activities in Washington. Telegraph messages, letters, and telephone calls assailed Congress by the thousands, all at the urging of the American Legion. Senator Johnson of Colorado expressed his dismay on learning of the state of dischargees: "I never realized that anything approaching this situation existed."[5] Echoing the drafters of the Declaration of Independence, and the leaders of the women's movement in Seneca Falls, advocates settled on a name for the omnibus legislation that would roll a wide range of benefits into one law: "the G.I. Bill of Rights," which would ultimately be shortened to the "G.I. Bill."[6]

Deploying a multimedia blitz, the Legion utilized newspapers, radio, and movie houses to promote the legislation. They charted the sentiments of individual members of Congress. If a member of Congress seemed to waver, telegrams would go out from the Legion in Washington to local chapters to apply pressure to win the vote.[7] Because the Legion feared a program for returning veterans that looked like a handout would stigmatize the beneficiaries, it worked hard to promote the message that the benefits available through the program were a debt to be paid to veterans, not a form of charity. When opponents of the legislation expressed fear that the benefits afforded through the bill would promote laziness, the *Army Times* responded with a blistering editorial stating emphatically that the G.I. Bill was not "charity." Instead, as the paper argued, it "merely provides a chance for service men and women to navigate under their own power." Service members had been "taken from jobs, homes and futures to win a war. Certainly they deserve a little assistance in making their readjustments."[8]

As crafted, the G.I. Bill was generous and inclusive, making few distinctions between service members or the extent to which they did or did not see combat. The Legion seemed to know something intuitively about public support for social programs. Programs that are inclusive, and have no income tests, are more popular and are accepted more widely. Historically, the U.S. public generally treats programs with broad eligibility criteria—like Social Security—favorably.[9] The last thing the Legion wanted to do was stigmatize returning veterans, and the broad and inclusive criteria for benefits were one way to avoid doing that.[10]

What is more, the American Legion offered an optimistic vision of national unity and economic strength that a broad-based program for returning veterans would foster. Its leaders stressed the fact that returning veterans had fought to challenge an existential threat to the United States, and that the broader populace should support them by making their reentry into the economy and local communities as easy as possible. Just as waging the war was a collective effort, rejuvenating the economy in the wake of the war was also something that required a national endeavor, calling for the cooperation of and shared sacrifice from everyone.

Similarly, when advocating for passage of the G.I. Bill, the Legion stressed the need for a broad-based program, one that would not draw distinctions, single certain veterans out for benefits, or leave others to go without. It worked to craft a broad-based program and later advocated for it, emphasizing the idea that many returning veterans, regardless of whether they saw combat or the length of their tour of duty, would receive benefits under the G.I. Bill. As a result, most returning veterans received some form of benefits under the program; in other words, it did not make distinctions among veterans, stressing their similarities, not their differences. The Legion also focused on the economic hardship that could befall returning veterans as well as the economic consequences to the economy as a whole if veterans could not reintegrate into economic and community life. The legislation's benefits would have a positive impact on not only returning veterans but on both the economy and society as well, and that was one of the optimistic messages the Legion tried to promote.

The American Legion's deft use of an optimistic, positive, broad-based message that attempted to personalize and humanize the problems returning veterans were experiencing led to its successful campaign to win passage of the G.I. Bill. Such messages tap into some of the trust-generating ideas discussed in the last chapter. First, such messages attempt to trigger the notion of homophily described earlier by drawing connections between people, helping them to see the ways in which they are alike, the ways in which they might be able to identify with the plight that others are experiencing. This has the effect of lowering social distance between people and groups of people. Second, by promoting inclusive messages, they seek network effects by growing a wide range of supporters. In many ways, the messaging itself creates, builds on, and expands a network of supporters. Ideas infiltrate formal organizations. Their members might support the effort, and then a social movement forms when such organizations work together. Successful social movements attempt to draw connections between disparate groups so that they will agree to work together toward a common goal. As we have seen, the stronger and broader the network, or the network of networks, the greater the network effects it can generate, building on the ties, both close and extended, that hold the network together.

Fast-forward almost twenty years from the passage of the G.I. Bill, and just months before his death, civil rights leader Medgar Evers gave the televised speech discussed in chapter 1. Although there are more famous speeches of the civil rights era that embraced positive, inclusive messages, with King's "I Have a Dream" speech being the most famous, the Evers address serves as one of the clearest examples of an attempt to lower the social distance between those in his movement and those he might convince to support it, let alone those opposing it. In this address, Evers appealed to a sense of common history and shared destiny between the black and white communities of Birmingham. He stressed the role that local individuals were playing in the organizing taking place in that city.

He emphasized the similarities and the personal connections between the parties on either side of the police batons. "I was educated in Mississippi schools," Edgars said, "and served overseas in our nation's armed forces in the war against Hitlerism and Fascism." He continued: "I mention this because I believe I am typical of many loyal Mississippians of color, who are equally devoted to their State and want only to see it assume its rightful place in the democratic scheme of our country." Evers explained that he understood that the NAACP, to most southern whites, "whether they are friendly or hostile," is a "'Northern outside group.'" He protested that the "facts do not bear this out." He emphasized that at least "one-half of the NAACP membership is in the South" and that there had been branches (i.e., translocal outposts) in Mississippi since 1918. "Therefore," he continued, "when we talk of the NAACP we are also talking about fellow Mississippians, local home-grown Negro citizens, born and reared in communities such as Jackson." Evers tried to draw from those things the black and white communities of Mississippi shared to try to lower the psychological barriers and perceived differences between those communities in an attempt to foster empathy and support for the cause of civil rights.

The striking nature of the violence against civil rights activists, particularly in the 1960s, when it was carried, sometimes live, over the television, garnered empathy for the victims of this violence. In order to achieve progress on civil rights, and serve a range of political goals, President Johnson, like Evers, tried to tap into the polity's sense of shared humanity and shared destiny when promoting legislative changes to protect such rights. Days after "Bloody Sunday," when a young John Lewis and other civil rights leaders were attacked while attempting to cross the Edmund Pettus Bridge in Selma, Alabama, President Johnson implored Congress—indeed, the entire nation—to support voting rights legislation that would outlaw processes imposed in the South that were designed to deny African Americans the right to vote. He did so by invoking notions of equality, shared humanity, and shared destiny. Johnson tried to appeal to a broad cross-section of the public and he appropriated religious imagery and language from the civil rights movement to advance the cause of voting rights legislation. For Johnson, the issue of equal rights for African Americans was a question that "lay bare the secret heart of America itself,"[11] and one Americans needed to address: "Should we defeat every enemy, should we double our wealth and conquer the stars, and still be unequal to this issue, then we will have failed as a people and as a nation." Quoting scripture, Johnson continued: "With a country as with a person, 'what is a man profited, if he shall gain the whole world, and lose his own soul?'"[12]

Solving the problem of racial inequality, Johnson argued, was not a "Negro problem," a "Southern problem," or a "Northern problem." It was "an American problem." He continued: "And we are met here tonight as Americans—not as

Democrats or Republicans—we are met here as Americans to solve that problem." He implored the American people to realize that "it is not just Negroes, but really it is all of us, who must overcome the crippling legacy of bigotry and injustice." Drawing from the language of the civil rights movement, he added, "And we shall overcome." In these ways, Johnson attempted to appeal to the self-interest of white Americans in his effort to garner support for voting rights legislation. For Johnson, the reason that Americans needed to come together to solve this civil rights issue was that "Negroes are not the only victims." Fighting equal rights had consequences for white Americans as well. "How many white children have gone uneducated?" he asked. "How many white families have lived in stark poverty, how many white lives have been scarred by fear, because we have wasted our energy and our substance to maintain the barriers of hatred and terror?"[13] In the months following this speech, Congress debated voting rights legislation and ultimately passed it in the summer of 1965.

As the Legion's use of individual stories of returning veterans, Evers's address, and Johnson's imagery in promoting voting rights legislation show, messages of inclusion, shared destiny, and shared humanity help personalize a cause and decrease social distance, all in an effort to activate and build broad support, rally allies, and strengthen a movement. Just as translocal networks foster cooperation and trust, certain types of messages draw in supporters and build a movement's size and strength. The messages that tend to do so are those that help individuals identify with the goals of a social movement, goals that may be in those individuals' interest to support for a host of reasons: those goals may serve to benefit those individuals or represent the type of change they want to see in the world, even if they enjoy no direct benefits from them. Tocqueville argued that when individuals engaged in collaborative, cooperative work with their neighbors, they were advancing their "self-interest well understood."[14] An individual's belief that he or she has something to gain from participation in and support for a social movement will encourage her to join in and engage in collective action geared toward social change. But is this idea simply one of self-interest or is it something more? As Tocqueville said, is it self-interest "well understood"? Moreover, does it go to the very nature of the self as well—the ways in which we see ourselves in the groups and movements with which we associate? The last chapter focused on the networks that make up one's store of social capital, one's networked trust. But what is it about a particular network that might attract us to join it?

For an understanding of some of the additional components of social movement success, particularly in social innovation moments, I turn to contemporary social movement theory to try to identify the connection between one's network, the messages that network might send, and the extent to which the identities of the members of that network are tied up in both. In this analysis, I bring together some of the concepts discussed in the first two chapters and weave in a third: the role that messaging plays in social movement success. I also discuss the relationship of that messaging to technology and organizational structure to understand the role they all play in fostering successful social mobilization. While early social movement theory stressed that movement leaders should focus on a sort of cost-benefit analysis when it comes to community organizing, later theorists focused on identity formation, which has at its core the messages a movement adopts and the relation those messages have to the members who join it. In what follows, I discuss this evolution of social movement theory, beginning with what I will call the rational actor model of community organizing. I then move to a focus on what critiques of this model say about the role that identity formation, social networks, and the messages that tie them together play in social movement success. What this discussion shows is that messages matter for community organizing and social mobilization. Personalizing, humanizing, and optimistic messages can help movements expand and grow, creating the network effects described in the last chapter. At the same time, when those messages are encoded onto our face-to-face relationships, those relationships serve as a channel through which a movement can expand its network. When it comes to social movements, the messages they send and the networks on which those messages are conveyed are deeply entwined. This understanding of a movement's message, and all that message entails, constitutes the third and final component of the social change matrix.

What Social Movement Theory Says about Networks, Identities, and Messages

Mancur Olson focused on the perceived costs and benefits in collective action, and this focus led to the development of what has come to be known as resource mobilization theory, which posited that individuals engage in a rational, cost-benefit analysis to assess whether, and to what extent, they might engage with a social movement in pursuit of collective action.[15] To borrow from economists who talk about the concept of "transaction costs,"[16] the costs associated with participating in and engaging with social movements are what the political scientist Sidney Tarrow calls social transaction costs: What will participation mean for the activist in terms of how much time she will have to spend, how much effort will be required, and what sort of economic loss she can expect?[17] With this rational actor model of organizing, the challenge for leaders of social movements is, first, to communicate to members and potential members the message that the benefits of activism will outweigh any burdens the activist will have to bear, and, second, to build efficient organizations that can maximize such benefits and reduce the costs of mobilization.[18]

I saw some of these dynamics play out as a lawyer for tenant organizing groups. In that context, I would discuss with many of my clients and potential clients whether it made sense for them to participate in the work of a particular tenant association. These discussions often led to questions about whether, if we took the landlord to court, a tenant association member would have to miss a day of work (and pay) to come to court to support the effort. Others would have to grapple with getting childcare for tenant meetings. Most tenants weighed these costs against what they hoped to achieve through participation. Some might seek better living conditions for their families. Others wanted lower rents. Still others, whose situations might be better off than some of their neighbors, wanted improved relations with their neighbors and to build a store of goodwill should they need help from their neighbors in the future. But the social mobilization calculus usually went beyond a simple weighing of costs and benefits. For example, some tenants might have been more inclined toward conflict with the landlord because of their own personality or poor experiences with this landlord or a prior landlord, or they might view the activity of the group through the lens of racial, ethnic or economic justice. Because of these experiences and perspectives, they might seek out opportunities through which they might find class, racial, or ethnic solidarity with their neighbors. Regardless of the benefits a particular tenant sought to achieve through the group work, such benefits, sometimes explicitly and consciously, sometimes implicitly and subconsciously, would often be weighed against the likely costs of participation. Thus my experience engaging in community organizing and working with many community organizers over the years tells me that the struggle over cost and benefits is a real one, particularly for those on the lower end of the income scale. Making matters worse in many situations, participation in a group's efforts can put what little economic security one possesses at risk. Engaging in union organizing can lead to a loss of shifts or of one's job altogether (which is, of course, supposed to be illegal). A tenant engaged in tenant organizing can face eviction over trumped-up charges. While it may make some sense intuitively though, the emphasis of this theory on rational, cost-benefit analysis, careful planning, and efficient organizational structures fails to account for all the ways in which effective social change comes about. Some degree of cost-benefit analysis may take place, but this does not tell the whole social change story.

Individuals do not make decisions about whether to engage in a social movement or not based solely on a cold, calculating interest in the desired ends of a social movement.[19] Critics of this rational actor model would say that any such calculation, if it happens at all, is tied to one's identity, experiences, sense of self, and any affinity one feels for and the extent one identifies with a social movement.[20] That calculation is not merely one driven by self-interest, as the rational actor model presumes.[21] Tied up in this notion of self-interest is not just one's assessment of the benefits that accrue from cooperation; it can also serve as a reflection of one's sense of self, one's identity. Participation in a group conveys a message about the individual members of that group: what they believe and what they want the world to believe about them. Each of these components of a group's message is tied up in the goals that group may have for the social change it pursues. These components also inform each individual member of the group's own sense of self: what she believes about herself and what she wants others to believe about her. Marshall Ganz's study of the California farm workers' movement in the 1960s revealed these complex components of the message-making aspect of group affinity, recruitment, and allegiance. Where rival groups—the National Farm Worker Association (NFWA) and the Teamsters—tried to organize the workers, which group an individual worker might join revealed something about the identity the worker wanted to embrace: "signing an NFWA card meant taking a risk, expressing solidarity with one's fellows, making a claim, and asserting an ethnic identity. Signing a Teamster card meant protecting one's job, doing what the boss wanted, and, in the eyes of many, denying one's ethnic identity."[22] As Ganz explained, "Signing an NFWA card expressed anger or hopefulness; signing a Teamster card expressed fear or resignation."[23] In other words, that choice both represented the embrace of an identity by an individual and a group but also sent a particular message to others.

In social movement scholarship, these notions of identities and messages are often incorporated in what are called "frames": a way one makes sense of the world and ultimately what guides one's actions. These frames are mediated through and embedded in the groups with which one is affiliated.[24] Similarly "collective action frames" help a group explain the injustice in the world, place blame for it, and offer a solution to it that requires taking responsibility for that change; when this occurs, it spurs action.[25] So called "frame alignment" arises when one's view of the world is consistent with that of a social movement.[26] What it means to be a part of that movement, the change that movement strives to achieve, and what participation in that movement says about those who join it are all tied up in the concept of a group's frame or frames. In turn, these frames are often embedded in and encoded onto the organizations and networks that adopt them.[27]

In the last chapter, I called networks that combined both bonding and bridging social capital "hybrid networks." As a network grows, its strength can also grow as that network develops and can generate network effects with greater and greater force. In the end, broader, more inclusive messages attract a broader cross-section of society because more people can identify with the message. The late Derrick Bell developed a theory of constitutional interpretation that many have taken as a theory of social change itself, one driven by those moments when there is a convergence of interests among groups within society as reflected in the messages they send.

When Interests Converge

Before becoming the first tenured African American law professor at Harvard, Bell was a lawyer in the U.S. Department of Justice and then for the NAACP Legal Defense and Educational Fund. In his legal and scholarly work, he advocated for an appreciation for the tortured history of race in the United States, as well as an understanding of how social change happens. His theories may help explain why certain messages tend to result in successful social change.

Efforts to promote racial justice, from the era of abolition to the civil rights movement of the 1950s and 1960s, have always centered on a particular message: equal rights and dignity for African Americans. Proponents of racial equality often stressed notions of shared interests, shared humanity, and shared destiny between communities of color and the white community. In addition, efforts to promote greater racial equality often used images and stories to personalize the experiences of African Americans, creating bonds that lowered the distance between people of color and the white community. What is more, images of racial violence in the South in the 1950s and 1960s were used as propaganda by leaders in the Soviet Union to cast the U.S. capitalist system as racist and unequal. This was a particularly powerful message in former colonial nations in Africa. When political leaders in the U.S. saw the need to promote greater racial equality in order to help combat the international public relations nightmare they were facing as a result of the Jim Crow system, the interests of national political elites converged with the interests of leaders in the civil rights movement. Remember President Johnson's conversation with Martin Luther King, Jr., referenced in chapter 1? There, in addition to talking about the importance of telling stories through the media, Johnson also said the Civil Rights Act of 1964 was the most important *foreign policy* achievement of his administration.[28]

Derrick Bell coined the term "interest convergence" to reflect the notion of how social change happens. And he found that it was one of the main drivers of the Supreme Court when it issued its landmark school desegregation ruling in *Brown v. Board of Education*. For Bell, the decision in *Brown* is at least partially explained by the interests of white policymakers who were "able to see the economic and political advances at home and abroad that would follow abandonment of segregation."[29] The decision provided "immediate credibility to America's struggle with Communist countries to win the hearts and minds of emerging third world peoples."[30] It would also help at home, where, Bell argued, service members returning from fighting in WWII saw that the Soviet Union had improved racial dynamics in the Soviet bloc in a single generation; that realization might have led to greater racial unrest on the home front. Finally, for whites who wanted to see the South progress economically and speed its transition from an agrarian past, the lasting vestiges of Jim Crow would hinder that economic progress.[31] For Bell, while notions of morality might have been enough to motivate many whites to support the cause of racial equality, "as with abolition, the number who would act on morality alone was insufficient to bring about the desired racial reform."[32] By tying the interests of disparate groups together, it helped build a groundswell of support for the change that would ultimately come about. Instances of interest convergence are not reserved simply for the civil rights era, however.

Returning to the effort to end slavery in the United States, this notion of interest convergence is laid bare in the early days of the Civil War, a century before Bell coined the term. In a letter to Horace Greeley, President Lincoln stressed that his "paramount" goal was "to save the Union," and that to do so, he believed he needed to end slavery. In this way, his war aims, and those of the abolitionists who sought to end slavery, would converge.

> *My paramount object in this struggle is to save the Union, and not either to save or destroy slavery.*
>
> If I could save the Union without freeing any slave, I would do it; if I could save it by freeing all the slaves, I would do it; and if I could save it by freeing some and leaving others alone, I would also do that.
>
> What I do about slavery and the coloured race, I do because I believe it helps to save the Union; and what I forbear, I forbear because I do not believe it would help to save the Union.[33]

Strong messages in support of racial justice have not just focused on equality and interest convergence, however. They have also attempted to link divergent communities by stressing personal connections between them, as Evers tried to do with his televised address. When social movements attempt to draw connections between seemingly disparate communities by telling poignant stories that stress the personal connection between individuals and communities despite what might otherwise be perceived as interpersonal differences, they stand a better chance of strengthening a movement by broadening its network of supporters. Like in the translocal, networked groups described in the last chapter, where repeated interactions and opportunities to cooperate can be driven by low social distance but can also reduce social distance, in ways large and small, movements can communicate messages that help build solidarity, lower social distance, help individuals identify with the movement, generate network effects, and foster cooperation and the emergence of social capital.

When I was a community organizer and lawyer working with groups seeking to expand their membership base, current group members and leaders would often tell their own personal stories in order to decrease the social distance between current and prospective members of the group in an effort to gain support and enlist new members. When working in Chinatown, representing restaurant workers there in campaigns for fair wages, I could see natural divisions between workers in a particular restaurant. The food service industry typically divides its workers into classes of workers, labeling them "front of the house" (waitstaff, bartenders) and "back of the house" (cooks, dishwashers, etc.). As someone who washed dishes as one of my first jobs as a teenager, I can tell you that these divisions often take on a class-based tinge: the front-of-the-house workers can, at times, look down on the kitchen staff. It may not be the strict social order portrayed in *Downton Abbey* between butlers, footmen, cooks, and maids, but the differences between these groups often permeate the culture in the food service industry, and these differences are often reflected in unequal compensation, perks, working conditions, and treatment.

In order to bring the groups together in ways that harnessed where their interests might converge, the leaders might tell their own personal stories of mistreatment by the restaurant owners to try to build personal bonds between the otherwise divided classes within the restaurants. If the workers all experienced unfair treatment from the owner, which would be conveyed to prospective members through personal stories relating instances of such poor treatment, perhaps the otherwise divided groups could see that they shared a common interest in pursuing changes in their working conditions. Shared interests could lead to shared action, as the ultimate success of the California tax revolt in the 1970s showed. In that setting, when groups that were not natural allies—like owners of large businesses in cities and suburban homeowners—teamed up, the California tax revolt of the 1970s succeeded.[34] In my own work and experiences as an organizer and lawyer, an appreciation for shared interests often flowed from the stories the group leaders told to those they wanted to convince to join the fold: stories that stressed, in a very personal way, how they were similar and downplayed their differences. Such personalized messages have proven an effective channel through which to bring in support, lower social distance, and prompt lasting change.

On a small scale, I saw this time and time again in my own organizing work. When I worked in Upper Manhattan in the 1990s as a tenant organizer and lawyer for community-based organizations, many groups would start small, and their members were mostly homogeneous in terms of race or ethnicity. The group might be made up primarily of African Americans or members of the Latinx community, sometimes from a particular country, like the Dominican Republic or Mexico. Their ability to organize into a tight-knit group at the outset might have depended on this homogeneity, this homophily. The group members would invite me into their homes to talk about the issues they wanted to address. Often it was a problem on their street, perhaps with local drug dealers, or their landlord who was not providing adequate heat during the winter or was failing to make timely repairs in their homes, leading to unsafe conditions for them and their families. The members of the group would want to take action and apply pressure on someone who had the power to make change—maybe it was their landlord, maybe it was a local elected official, or maybe it was the local police precinct commander. Invariably, we would talk about trying to increase the group size, knowing that a larger group would signal to the decision maker the group was trying to influence that he or she had to listen and take action in response to the group's advocacy. But growing the group would mean diversifying, moving it from a homogeneous group to one that was a better cross-section of the immediate community, whether it was within their building, their block, or their neighborhood. This would require an appreciation for the interests and needs of a broader coalition of individuals and families from the pool of potential members from which the group could draw support. This often involved reaching out, one by one, and face-to-face, to those individuals who might support the effort in order to understand their interests and needs. The existing group members would try to find where the interests of current members might overlap with the interests of potential members and look for common ground. And the messages that resonated with potential members typically involved the presentation of a hopeful, optimistic vision of what was possible if those potential members joined the group and everyone agreed to work together. By working together, the group members could achieve some positive good for the community, a chance at improvements in the lives of the members and their families. These types of messages are the ones that would rally supporters to them and offer a positive vision of the benefits of group activities. But there also has to be a sense of efficacy: a positive vision that the work and sacrifice that group activism entails will lead to the change the group wants to see. Thus tied up in the group's ability to motivate members and potential members is the ability to convey a message that resonates, galvanizes, and motivates. And, it would seem, there may be a science to which messages "stick" and which do not.

The Biases That Build Movements and Their Messages

To return to critiques of the rational actor model of community organizing, an additional flaw of this view is that it does not account for many of the ways in which we are not always rational in our decision making. We have learned enough about the functioning of the human mind and decision making in recent decades to know that our perceptions of situations and the decisions about the likely success of actions in the future in such situations are inherently flawed by a range of biases.[35] In social movement settings, social change activities are not easily reducible to a formula or algorithm, particularly when the individual making the calculation has inherent flaws in her subjective perceptions of a range of variables that would go into that calculation and values certain benefits differently than someone else.

In addition, critics of the rational actor model of organizing argue that individuals decide to join social movements for many reasons and those reasons are not always simply to maximize one's own, personal benefits. Each individual is different, with different reasons for engaging in social movement efforts, different levels of tolerance for difficult tasks, and differences in one's view of self-interest itself.[36] One individual's past experiences with civic engagement and the social meaning one gives to that experience—was such work successful and was the sacrifice worth it?—will also impact one's willingness to engage in future activity.[37] At the same time, there may be others who, despite long odds or even no chance of success, will still want to participate in a movement. In such situations, their actions send a statement, not only to those whose behavior they want to change through their efforts, but also to the other members of the group, and even, in a way, to themselves. But are there ways that organizers can harness some of the biases that might operate to inspire action, and avoid those that tend to discourage it, to advance social change? The groups highlighted here seem to have crafted messages that play to the biases that tend to inspire, to great effect.

Optimistic messages of unity and greater equality that build on personal connections seem common in successful social movements that emerge in social innovation moments. We can see these messages in the American Legion's efforts to promote the G.I. Bill and that of the civil rights movement of the 1960s. Why do these messages appear to have sparked some of the successes I have highlighted so far? The field of behavioral science—which itself learns a lot from game theory—offers some insights into why it is that these types of messages, when paired with an effective medium and a broad-based, translocal network, have often been at the center of many successful social movements that have emerged over the course of U.S. history in the wake of the introduction of new modes of communication.

Behavioral science tells us that two biases—optimism bias and confirmation bias—may be at play when talking about what makes optimistic, positive messages stick and catch on. *Optimism bias* refers to our common presumption that bad things will happen to others and not ourselves and that the future looks rosier than the present or the past.[38] While there are certainly individual, cultural, and national differences that tend to influence the degree to which a particular person may be optimistic, study after study tends to show that the human condition may, generally, be an optimistic one. This is particularly true where we think we may have control over a particular outcome, or we can create social distance between ourselves and a person we may deem as more stereotypically likely to have something happen to him or her.[39]

Another bias, *confirmation bias,* refers to our tendency to accept information that confirms previous beliefs as opposed to that which might challenge them. We also seek out sources of information that play to our confirmation bias.[40] It is no secret that individuals from different political perspectives tend to gravitate toward sources of information that confirm their preconceived notions. Liberals turn to Rachel Maddow and MSNBC, and conservatives to Fox News and Rush Limbaugh. It is far easier to let in information that confirms preexisting beliefs than to challenge those beliefs by identifying their weaknesses. As Drew Westen, a political consultant and psychologist, explains, "Our brains have a remarkable capacity to find their way toward convenient truths—even if they're not all that true."[41]

Based on these two biases, humans bend toward optimism, and when we hear optimistic messages, these tend to confirm this preexisting tendency toward optimism. Is it any surprise, then, that optimistic messages, ones that stress unity and greater equality, are those that, history has shown, are the ones that might generally attract wider support than ones based on division and negativity? Social movements should see some value in sharing personalizing messages that lower social distance, communicate an optimistic vision for the future, and often evoke a sense of a common and shared destiny and shared humanity. These are the themes that many of the social movements described here have attempted to promote. By striking these types of themes—by lowering social distance and building on an optimistic sense of shared destiny and shared humanity—social movements, to the extent they are able to rally supporters, seem to connect with something engrained in the human experience.

But can a larger, non-networked movement still activate the likelihood that an individual will identify with it? Face-to-face interactions and relationships were often at the heart of translocal, network organizing, as described in the last chapter. Research into why individuals join social movements indicates that it is one's ties to one's friends and close, preexisting relationships, and the trust one places in those ties, that is often more important than ideology to mobilization. One study of participants in Freedom Summer during the height of the civil rights movement showed that, while ideology certainly was a pull, drawing people into participation, the difference between those who joined and stayed with the group and those who dropped out of it was the strength of the former group's preexisting relationships to participants in the group.[42] Similarly, using various methods, including a metastudy of research on outreach methods used by groups as well as qualitative interviews with members of religious organizations and college students, one study found that an individual's preexisting relationships with members of a group was a stronger indicator of whether that individual would join such a group than whether his or her ideology aligned with the group.[43] As Tarrow explains: "Although it is individuals who decide whether or not to take up collective action, it is in their face-to-face groups, their social networks and their institutions that collective action is most often activated and sustained."[44] If mobilization is a product of face-to-face interactions with individuals to whom we have a preexisting relationship, it is no wonder that translocal networks are more powerful engines for the launching of social movements. Small translocal groups help lower social distance, and messages that share these personalizing, optimistic features are easily spread in face-to-face meetings, where they can also lower social distance. And it is these types of meetings that historically were common in the translocal organizations that formed the traditional networks through which social change pulsed for roughly two hundred years. These face-to-face interactions also activate another cognitive bias: that of *availability*. People tend to make judgments based on the information most readily available to them, the concepts that are most prevalent in their daily lives. When we receive messages from our closest relations who are members of particular groups, we might come to the conclusion that a particular message is highly popular and broadly accepted.[45]

Once an individual feels a connection to a movement's message, it is easier to activate her engagement with the work of the group. But that individual probably needs some personal connection to the group as well, some way into that group. An individual's connections to social networks help shape not just that individual's way of looking at the world; they shape her identity itself and impact that individual's willingness to engage in advocacy with the group. Those connections serve as a springboard, platform, and channel for mobilization. The extent to which an individual is embedded in existing social networks can determine not just whether one will identify with a particular social movement, but also whether one will engage with it.[46] The task for a movement is to activate these different components of the motivation calculus. As the sociologist William Gamson argues, identity, calculation, and mobilization are intertwined, and groups "that have achieved a successful integration of personal and collective identity will have an easier time doing what it takes to launch many kinds of collective action."[47]

Thus ideology and identity can meld with mobilization strategies by connecting an individual to the organizing efforts of groups with which she identifies and the messages those groups send. What is more, the preexisting connections one enjoys to the individuals in such networks and the sense of self one might see reflected in one's commitments to such networks and the messages they espouse can spark movement participation and activism. Thus one's preexisting social relationships are at the heart of community mobilization: they serve as both a source of efficient mobilization but also reflect our identity, our sense of self.[48] Examples of this connection between an individual, his or her preexisting social network, and the growth of new social movements out of such previously existing relationships abound. The civil rights movement got its start and found its base in African American churches.[49] The women's movement in the 1960s grew out of preexisting community networks.[50] Furthermore, it is the ability to activate such preexisting networks that can often make the difference between the success or failure of a social movement.[51] For example, although it was not exactly a progressive social movement, the nascent tax revolt movement in California in the mid-twentieth century showed an inability to tap into preexisting social networks, which would ultimately lead to its initial demise. Their first attempts at advocacy would fail because the grassroots leaders of the movement—homeowners from the suburbs surrounding Los Angeles—did not have a preexisting network to co-opt for support; did not inspire support from potential allies with aligned interests, like those in the business community in the city that did have such a network; and did not have a means of communication across the different suburban geographies that ringed the city.[52] Similarly, a comparison of the growth of the pro-life and pro-choice movements in the 1980s found that the relative success of the former in generating engagement of its members when compared to the latter was largely a function of the pro-life movement's ability to mobilize members from preexisting networks like faith-based groups.[53] Other research into such movements made similar findings with respect to how those groups were first started and how they initiated recruitment among new members: that is, prior relationships among individual members through preexisting networks were the primary recruitment tool as compared to other groups that relied on professional organizers with no previous attachment to the individuals they were trying to recruit.[54]

In these ways, a group's message is wrapped up in the identity of individual members and the ties they feel to the collective identity of the group as a whole.[55] It is not difficult to see that we are more prone to feel a connection to those with whom we already identify and it is these connections that lend us our identity. What is more, these connections are the wellspring of trust and can be tapped effectively for recruitment purposes. Here we see some of the network science aspects of social movements take hold: the more one can identify with other members of a movement, which represents the application of homophily (similarity) and propinquity (proximity), the more likely it is that one will participate in collective action to help strengthen and build the network that makes up the movement. The network thus draws a participant into it through that individual's identification with the individuals as well as the values of the network. That network shapes and is shaped by the identities of those who participate in it, and those identities are tied up in the message or messages a group embraces.[56] As an example, one study of the lesbian feminist movement in the United States found individual and collective identity were both essential to that movement's mobilization.[57] This notion of shared identity is also closely tied to notions of shared destiny. The more alike people see themselves, the easier it is to see they share a common fate.[58]

But how does this translate into a movement? Similar to the notion of bonding and bridging social capital, a group promoting a message that may tie a smaller group together can gain strength when it links its message to those of other groups. When this occurs, the phenomenon of "frame bridging" occurs, when different frames embraced by different groups mesh to create a single frame: a way of looking at the world that enables cooperation and the typing of different networks together.[59] And the broader this alignment and bridging, consistent with the interest convergence thesis, the larger the movement and the greater the network effects that flow within and from it. Throughout history, the ability to create such frame alignment and frame bridging has been a critical source of social movement success. Citing the early Christian movement in Roman times, the Indian independence movement, and the American civil rights movement of the 1960s as examples of "cultural hybrid[s]," Doug McAdam has argued that "those movements that have been especially important as sources of cultural innovation would seem to be those that resulted in meaningful, that is, egalitarian, contact between previously segregated social strata."[60] In these hybrid groups, we see contact between disparate groups, a recognition of shared interests, and then a willingness to collaborate and cooperate toward a collective goal that is broadly shared.[61] This last part—that a movement needs to tap into feelings of solidarity and identity that may already exist—is consistent with the previous discussion about movements building on preexisting, face-to-face networks. It is in those networks that these feelings of solidarity and identity often thrive.

Thus it would appear that there is deep connection between the messages a group sends and the networks, ideology, and identity that help trigger social activism. At the same time, certain types of networks and preexisting groups facilitate not just strong identity formation but also mobilization itself: that is, groups that foster face-to-face contact and lower social distance make recruitment more likely.[62] Individuals may identify with the cause a social movement espouses, but it takes their previously existing social connections, their ties to the network formed through interpersonal contact, that foster recruitment and then participation. And the type of network structure that is more conducive to this type of recruitment is that which occurs in the small, local, grassroots groups built on face-to-face interactions described throughout this book. What then converts them to translocal networks is the ability of the smaller groups to identify with the greater cause around which the larger organization is mobilized. Thus messages, networks, ideology and organization are all intertwined. It is not "'ideology' versus 'organization,'" as Tarrow explains; rather, these are "complementary solutions to the problem that movements need to solve."[63] In these ways, translocal network organizing may help bring together the somewhat conflicting accounts of social movement success: the rational actor model and those theories based more on identity formation. That is, we find in our preexisting relationships, which are embedded in the translocal networks of which we are a part and that become the engines of social change, not just our identities, but also an efficient and effective means of mobilization: strategic organizing meets the search for identity.

As outlined earlier in chapter 2, at least three phenomena tend to make people more willing to cooperate: repeated, positive interactions; decreased social distance; and communication. These all tend to generate cooperative behavior. When put in the context of a sociology of social change, we can see why, historically, broad networks of translocal organizations may have proven successful in fostering cooperation and bringing about such change. Cooperation can flourish when individuals work in networks built on local, face-to-face interactions.[64] Such networks foster communication and allow individuals to build the trust necessary to work collaboratively toward a shared goal. But what gets individuals in the door of the union hall, community center, or a neighbor's living room in the first place? Here's where interests and ideology come into play, which brings resource mobilization theorists and their critics together: when one cannot just *identify* the interests he or she has at stake in the decision to support a movement but can also *identify with* both the message that movement espouses as well as the individuals he or she knows who are already present in the movement, the more likely it is that he or she will participate.[65]

What is more, when a group's message is one of shared humanity and destiny and attempts to personalize the message's appeal, it helps rally support from a broad cross-section of any particular community, sometimes even the nation as a whole. Getting individuals to work together is central to successful social change. Groups that are more broad-based and can rally a larger cross-section of the polity would seem to be those that promote an optimistic vision that combines an optimistic belief in the group's potential efficacy with a positive vision of the future that is possible. Legendary community organizer Saul Alinsky would describe his political philosophy as one "anchored in optimism." "It must be," he added, "for optimism brings with it hope, a future with a purpose, and therefore, a will to fight for a better world."[66]

Returning to the interplay between biases and messaging, if we are to believe that there is a connection between confirmation bias and optimism bias, it is those optimistic messages that are unconsciously more attractive and thus more likely to draw people toward them. In contrast, those movements that advocate a negative, pessimistic, and ugly message find it harder to create strong networks across social distance, weakening their ability to pursue their ends. So-called citizens' councils opposed the civil rights movement in the 1950s and 1960s. Unlike with Medgar Evers's positive, inclusive message, and those of other civil rights leaders like Martin Luther King, Jr., television interviews with the leaders of these councils—who often communicated racist invective—often cast them in a harsh light, light which likely turned off most Americans watching violence unfold across the South. White supremacist groups today echo those same, divisive, negative messages.

What these and the other examples used in this chapter reveal is that messaging matters. It can attract or repel supporters. It can help sustain members. It can foster goodwill and a willingness to continue the struggle. What is more, the nature of the message a social movement chooses to advance may have some impact on the movement's ultimate success because different types of messages can attract supporters, create interest convergences, and grow a network. A larger network made up of smaller cells—that combination of bonding and bridging social capital—can generate both the trust necessary to sustain a movement and the network effects that are needed for a movement to have impact. But messages are often embedded in the networks we join; these networks then become the source of not just our identity but also the initial mobilizing acts that launch social movements. We saw in chapter 1 that social movements that harness technology that assists groups in forming translocal groups can tap into the trust that is generated in such networks. We have seen so far in U.S. history, from the founding of the republic through the civil rights movement, that even when powerful technologies existed—like the television—that allowed one-to-many communications, the social movements that formed tended to organize themselves into translocal networks: federated, nested enterprises where face-to-face communications could still take place. All that would change, however, when a new type of technology emerged soon after the civil rights movement and achieved some of its greatest victories. This technology altered the structure of social movements for two generations, with profound impacts on U.S. associational life and social movements, as the next chapter shows.

4

THE GREAT DIVIDE

Richard Viguerie's political life has spanned decades, presidents, generations, and three epochs of political organizing. He has both gone against the grain and been at the forefront of mass movements. He began his political career in the late 1950s, during the Eisenhower administration. Soon after the election of John F. Kennedy, and at the beginning of the ascendancy of the liberal establishment of the 1960s, he held a leadership position in the conservative organization Young Americans for Freedom (YAF). He moved from Texas, where he was the chairman of the Harris County Young Republicans, to take a job working for the newly formed YAF, really working out of the marketing firm Marvin Liebman Associates in New York. A marketing expert by trade, who says he still reviews marketing research two or three hours a day, he used his skills to promote conservative causes. In 1964, he worked hard to get Barry Goldwater elected. At that time, direct marketing was still in its infancy, and using the technique was relatively rare in political campaigns. Moreover, identifying potential supporters was difficult.[1]

Working for Liebman, Viguerie was given the names of several large money donors and was asked to reach out to them, people like World War I ace Eddie Rickenbacker and Charles Edison. But, Viguerie soon learned, "I didn't like asking people for money; So I started writing letters." And he found he had a knack for it. The trick was, however, finding out to whom he should write.

Viguerie understood that one way to identify supporters of a particular candidate, short of door-to-door canvassing and laborious house-to-house telephone calls, was to go to a repository in the U.S. House of Representatives in Washington. There, a list of the names of donors who had made contributions of more than $50 to particular candidates was maintained. In late 1964, after the defeat of Barry Goldwater, Viguerie went to Congress and began to record the names and addresses of the donors listed there. In longhand, he wrote over twelve thousand, three-by-five inch index cards. Viguerie took these completed cards back to his office for input into a database. He worked with computer experts who had day jobs with a local charity that allowed them to use the organization's mainframe computer at night. After turning over the names he had recorded to these experts, one of them returned to Viguerie's office several weeks later with the database. Having only worked with computer punch cards in the past, when he was handed a magnetic tape on a reel, he unraveled it, not knowing what to do with it. He expected holes in the magnetic tape like the holes in punch cards. It is likely that this was one of the first computerized databases maintained by a political group.[2]

This database and others like it would revolutionize not just political but social movements for forty years. Pioneers like Viguerie, together with Ray Bliss and William Brock, two heads of the Republican National Committee (RNC) during the 1960s and 1970s, utilized direct mail tactics to connect individuals who might be sympathetic to conservative causes to the national groups and politicians seeking their support.[3] The goal was to give like-minded individuals information that would resonate with them and to convince them to vote for conservative candidates and donate to conservative causes and political campaigns. For Viguerie, there was a "blockage" in the press; conservative messages could not get through the mainstream media at the time. Conservatives needed a way to carry their messages right to the voters, without going through the media. What's more, direct marketing and direct mail, in addition to sharing a conservative message, could help bring in funding, get voters to the polls, and identify potential grassroots leaders.[4]

Although the RNC started a direct mail program in 1962, computers aided the effort later in the decade, and, by 1968 the RNC had raised over $6 million through this method.[5] The effort gained steam and was so successful that "by the late 1970s, direct mail was bringing in 75 percent of the RNC's receipts."[6] The late senator Daniel Patrick Moynihan lamented the success of the Republican Party in mastering direct mail and expressed fears that the Democrats were at risk of becoming a permanent minority party. What hit the Democrats, according to Moynihan, were "two things": "money, money, money, money . . . organization and technology."[7]

The ability to master this new technology came at a particularly propitious time in electoral politics. Indeed, Republicans were able to harness technology just as the electorate was changing dramatically: larger numbers of voters began to register as independents and the voting age was lowered.[8] In addition, in 1971, Congress passed the Federal Election Campaign Act, which imposed limits on the size of donations from individual donors. As a result, campaigns were forced to broaden their base of donors to get a greater number of smaller donations from a larger group of individuals. Direct mail was essential to such an effort. A computerized database of potential donors was an essential component of this kind of fundraising.[9]

In 2004, Viguerie and his coauthor David Franke wrote as follows:

> Since the beginning of the conservative movement, the U.S. mail has been conservatives' principal method of communication. Even today, conservatives may get most of their political news through talk radio, cable television and the Internet—an opportunity not enjoyed by previous generations of conservatives—but most activist communication between individual conservatives and their organizations and causes and candidates takes place through the U.S. mail. In time, we expect the Internet to become the foremost medium for activism, but at the present time that's still a work in progress.[10]

For Viguerie and Franke, direct mailing allowed conservative campaigns to "awaken the conservative within" of those who are identified as a target of mailing and received information from conservative sources, perhaps for the first time.[11] The direct marketing also did something else: it helped build a personal connection between a candidate or an institution and a potential voter or donor. Viguerie tells the story of Phil Crane, a conservative congressman from Illinois who won an election for the House seat left vacant when Donald Rumsfeld left the House to join the Nixon administration. Viguerie prepared a personal, "conversational" mailing that cast the candidate and his family in a soft, humanizing light. For years after that first election, which Crane won, people would walk up to the Congressman's wife and tell her that they remembered the mailing and how much of an impression it had made on them.[12] Politicians and their handlers across the political spectrum, at all levels of government since then, have replicated this approach in countless elections across the country, with a profound effect on political discourse and electoral politics. As the historian Jill Lepore explains: "Direct mail and cable television segmented the electorate and balkanized the public. Conservatives didn't waste their energy talking to voters outside the demographic they hoped to reach, which saved them money and made their campaigns more efficient; new technologies also provided candidates with an incentive for invective. Above all, they allowed conservatives to bypass the mass media, newspapers, and the gatekeepers of broadcast television, which, increasingly, conservatives represented as the enemy."[13]

By 2004, the Viguerie firm claimed to have mailed over 2 billion letters over the previous forty years, with over 100 million pieces of mail in the 2004 election cycle alone. That cycle, according to the estimate of Viguerie and Franke, involved over 3 billion mailings from political groups; 1.4 million "magazines, newsletters, and policy papers" from such groups; and 2.7 billion letters mailed by candidates at all levels of government.[14]

The ability to engage in mass mailing efforts figured prominently in the election of Ronald Reagan as president, and helped usher in the long arc of the Conservative movement from the early 1970s through the presidency of George W. Bush. But this ability did not just impact conservative causes. The victories in the cause of civil rights for the African American community spurred other movements as well: the environmental movement, the women's movement, and the gay rights movement, to name just a few. But these movements arose just as the ability to engage in targeted, mass mailing was coming into vogue. They looked at the victories of the Civil Rights Movement and were inspired. But, like many social movements before them, they also picked up the nearest and most advanced communications tool at their disposal—then the ability to generate mailing lists—and deployed it to create organizations that were quite different from the national, but translocal, organizations that helped advance the cause of civil rights in the 1950s and 1960s. As a result, unlike the Civil Rights Movement of those decades, which was built on networks of cells of grassroots groups spread out through the country and coordinated, loosely, by national organizations like the NAACP, these other movements, for the most part, utilized the ability to engage in mass mailing to create national organizations divorced from grassroots networks. Mass mailing would shape social movements for two generations and the next forty years. The rise of these technocratic organizations represented a sharp break from over nearly two centuries of social mobilization. Civic associations flourished and expanded their membership rolls in the two decades after WWII, just as they had in the wake of the Civil War. One study of the membership of thirty-two large, national organizations throughout the twentieth century showed a steady rise in the percentage of the population that were members of these groups (depending on the class of individuals who could be members in each group – for example, the percentage of youth who were members of 4H). This study revealed that apart from a decline during the Depression, these groups enjoyed a steady rise in membership throughout the first sixty years years of the twentieth century, and then a steady decline from the late 1960s to the end of the century. The steadiest and most consistent growth occurred during the post-WWII period.[15]

As Robert Putnam explains, this growth was a product of the U.S. public redirecting the "massive outpouring of patriotism and collective solidarity" that occurred during WWII into community life. From 1945 until the mid-1960s, those decades "witnessed one of the most vital periods of community involvement in American history." The "market share" of the overall population that joined these 32 large, national organizations would "skyrocket" during this period, and "[b]ecause of growing population, the increase was even more dramatic."[16] This "civic explosion" included older line organizations like the Grange, as well as relatively newer organizations, like the League of Women Voters.[17] According to Harvard University's Civic Engagement Project, in 1955, groups like the AFL-CIO swelled to over 12 million members, representing 12 percent of the adult population. Similarly, the National Congress of Parents and Teachers/PTA had over 9 million members in over 40,000 local chapters. The American Legion: nearly 2.8 million members in nearly 17,000 posts. Even the Independent Order of Odd Fellows, first formed in 1819, boasted half a million members in 7,500 local lodges.[18]

The civic pride in the efforts of the so-called "Greatest Generation" to overcome Depression and fascism led to greater civic participation and engagement with the political process. As political scientists Gabriel Almond and Sidney Verba's study of civic participation in the 1960s revealed, many of the groups of which individuals were members were engaged in political activities.[19] This combination of greater participation in the political process may have been one of the main drivers of an important phenomenon of this time: in the mid-1960s, roughly 75 percent of Americans thought one could trust the federal government in Washington to do the right thing all or most of the time.[20] Similarly, generalized trust in others was just about the same: 77 percent in 1964 said that one could trust others.[21]

Of all of these organizations, the groups that made up the Civil Rights Movement probably had the most profound effect on society, both directly, through their legislative, cultural, and judicial victories, but also more subtly, as they inspired other groups to action. Advocates on the left were learning lessons from these movements, and they would begin to form in the late 1960s and early 1970s. These groups would pick up on very particular lessons from the advocacy efforts of African American groups that had obtained landmark victories in the courts and Congress. But the lessons those groups chose to learn, spurred by the advent of the new technology of the computerized mailing list, would divert energy from the painstaking work of constructing trans-local networks designed to build political power and redirect it mostly towards the creation of national organizations with only a loose connection, if any, to grassroots movements. For many groups, it is possible they mistook victories for tactics, hoping to achieve the former, without recognizing the centrality of networked, trans-local organizing to those victories. As sociologists Pamela Marwell and Gerald Oliver argued, the new technologies appear to have taken a life of their own, becoming "virtual ends in themselves." [22]

Indeed, the Civil Rights Movement of the 1950s and 1960s was made up of several organizations that were trans-local, cross-class, and cross-race. From the Student Non-Violent Coordinating Committee and the Southern Christian Leadership Council to the National Association for the Advancement of Colored People, these organizations were made up of local chapters tied to a national network. Complementing these trans-local groups was the "LDF", the legal arm of the NAACP, which formed as a separate organization in 1940. Formally known as the NAACP Legal Defense and Education Fund, the LDF helped mount the successful legal challenges that ultimately led to the landmark decision in 1954 in *Brown v. Board of Education*, which made segregation in education illegal. The LDF would bring groundbreaking legal challenges to discrimination in housing, public accommodations, and institutions of higher education. Over the years, the leadership of the organization would include such legal lions as Thurgood Marshall, Constance Baker Motley, Jack Greenberg, and Julius Chambers.

Seeking to emulate these successes, identity- and issue-based groups formed in the wake of the victories of the Civil Rights Movement, many taking on the "legal defense and education fund" name, and, more importantly, its tactical focus. In 1972, advocates seeking to protect the rights of Puerto Rican Americans through litigation and the courts formed the Puerto Rican Legal Defense & Education Fund (PRLDEF) (now LatinoJustice PRLDEF). The Lambda Legal Defense and Education Fund (now Lambda Legal) got its start in 1973, in order to pursue and protect the legal rights of the lesbian and gay community through legal action. Similarly, environmental groups like the Environmental Defense Fund; the Sierra Club Legal Defense Fund (a spin-off of the Sierra Club, formed in the Progressive Era), which is now called EarthJustice Legal Fund; and the Natural Resources Defense Council all formed in the late 1960s and early 1970s to pursue legal remedies to preserve and protect the environment. As legal scholar Joel Handler argues, "[t]he apparent successes in civil rights litigation and the receptivity of the Supreme Court and the lower federal courts encouraged other groups and organizations to adopt a law-reform strategy."[23]

More conservative-leaning interests were also taking notice of the tactics deployed by Civil Rights groups and labor unions and began to consider ways to counter the attacks (as some saw them) on business interests, capitalism, and the free enterprise system. A powerful lawyer at the time, Louis Powell, who was involved in politics and assisting the tobacco industry in Virginia, and who would soon become a justice of the U.S. Supreme Court, penned a memo for his neighbor, who was working at the time for the U.S. Chamber of Commerce. The Powell memo, entitled "Attack on American Free Enterprise System," opens with a statement about what the author considered the "dimensions" of the attack.[24] "No thoughtful person," Powell asserts, "can question that the American economic system is under broad attack."[25] For Powell, the sources of this attack were "varied and diffused."[26] While they included "Communists, New Leftists and other revolutionaries who would destroy the entire system, both political and economic," these were but a "small minority" and not "the principal cause for concern."[27] For Powell, the "most disquieting voices joining the chorus of criticism, come from perfectly respectable elements of society, from the college campus, the pulpit, the media, the intellectual and literary journals, the arts and sciences, and from politicians."[28] Powell also blamed the media, particularly television, which "plays such a predominant role in shaping the thinking, attitudes and emotions of our people."[29]

The Powell memo then lays out what he says are the components of a strategy for business interests to defend themselves from the attacks on the system. Echoing trans-local movements in the past, Powell thought that such a strategy would start with the Chamber of Commerce, which had "hundreds of local [chapters] which can play a vital supportive role."[30] The notion that the Chamber had these local affiliate entities was, in Powell's words, "of immeasurable merit."[31] For inspiration, Powell turned to the examples of progressive organizations that could serve as models for the business community. "Lessons can be learned from organized labor," he argued.[32] He would write, "over many years the heads of national labor organizations have done what they were paid to do very effectively."[33] Although they "may not have been beloved . . . they have been respected—where it counts the most—by politicians, on the campus, and among the media."[34] Powell declared that it was "time for American business—which has demonstrated the greatest capacity in all history to produce and to influence consumer decisions—to apply their great talents vigorously to the preservation of the system itself."[35]

Many conservative groups would form, or take on new and expanded roles, around the time that Powell was urging business interests to find their voice, learn from social movements, and defend capitalism. One group, founded in 1943 as the American Enterprise Association, changed its name in the early 1960s, taking on the name the American Enterprise Institute (AEI). The group ran into some legal trouble soon thereafter as it was accused of improperly engaging in electoral politics, illegal for a non-profit like AEI, when members of its leadership supposedly lined up support for the candidacy of Sen. Barry Goldwater for president in 1964. But just as new rights groups were forming on the Left, wealthy donors were lining up support for new organizations on the Right, like AEI, that could serve as political and intellectual counterweights to more liberal organizations. William F. Buckley formed the American Conservative Union on the heels of

Goldwater's defeat by Johnson in 1964. The early 1970s, in addition to seeing the growth of such groups as Lambda and PRLDEF, also saw the formation of the Heritage Foundation, which was bankrolled by brewing magnate Joseph Coors and Richard Mellon Scaife.[36] Activists formed the National Taxpayers Union in 1969 to advocate for lower taxes. It now presses for a more streamlined and less burdensome tax system, reduced government spending, and a lessened regulatory burden on businesses.

Not to miss out on the trend towards the creation of organizations that would use the courts to pursue their political agenda, the conservative Pacific Legal Foundation was founded in 1973, and in 1977, the formation of the Washington Legal Foundation followed. Similarly, in 1973, conservative elected officials and advocates created the American Legislative Exchange Council (ALEC), which works with elected officials to promote the adoption of legislation at all levels of government that is favored by conservative, often pro-business, groups. Political organizations with an explicit religious bent also arose during this period, with the Moral Majority forming in the late 1970s, and then dissolving several years later, only to see the rise of the Christian Coalition in 1989.

While some of these organizations, like the National Taxpayers Union, are nominally membership-based (though it is difficult to discover on that group's website how to join as a member, but very clear how one could donate to the organization), the conservative groups that found new life in the 1960s, or were formed in the 1970s and beyond, mostly take the shape of many of the new groups on the Left: they are professionalized, non-membership based organizations with a central staff, seeking funding from individual donors (both wealthy and average citizens), businesses and foundations. They, like their left-leaning counterparts, eschew traditional, membership-based organizing of trans-local networks for national, issue-focused organizations without a grassroots base. It is no surprise then that this growth in legal organizations on the Left, and similar groups on the Right, coincided with an overall decline in the traditional, membership-based groups that had reached their peak in the 1950s.

Sociologist Theda Skopcol has documented the decline in these traditional organizations, noting the percentage change in membership from 1955 through 1995, a time when the overall population of the United States roughly doubled. Many groups saw their membership as a percentage share of the adult population drop considerably. The Free Masons saw a seventy percent reduction in their membership as a percentage of the adult male population. The Fraternal Order of Eagles, a forty-six percent drop. The Odd Fellows: ninety-two percent. The American Bowling Congress, which saw its membership swell by 136 percent from 1955–1965, experienced a total decline of twenty percent by 1995.[37]

Another area in which more traditional civic organizations saw significant declines during this period was in union membership. These working class organizations suffered during this period as well, as the percentage of non-agricultural workers that were unionized was cut in half, from one-third of the workforce, to one-sixth. Today, a little more than one-third of the public sector workforce is unionized, while fewer than one in fourteen private employees are in a union.[38]

Despite this decline in more traditional organizations, the late 1960s and early 1970s saw an explosion in groups supporting women's rights and the rights of a range of ethnic groups described above. As sociologist Debra Minkoff described in her research into the growth of civic organizations from the 1950s through the 1980s, the number of such organizations in the early 1950s numbered just under 100. By the 1980s, this number had increased over six times, to 688. Much of this growth was in groups advocating for the rights of women and ethnic minorities, and many of these organizations were what Minkoff calls "service" organizations—those not engaged in agitation outside traditional networks of power to bring about social change. During this tremendous growth in advocacy organizations, the number of groups involved in this political agitation—think of the Southern Non-Violence Coordinating Committee (what Minkoff calls "protest" organizations)—remained basically unchanged during this period.[39]

The battle over passage of the Equal Rights Amendment (ERA) would show the difference between old and new organizing strategies. In the pursuit of gender equality, the National Organization for Women (NOW) got its start in 1966. It attempted to assume the same sort of trans-local structure of the Civil Rights Movement and social change organizations throughout American history. Also mimicking the Civil Rights Movement, in 1970, it spun out a legal arm: the NOW Legal Defense and Education Fund (now Legal Momentum). NOW and its legal arm would go to work immediately to pursue a landmark change in the law: adding the ERA to the U.S. Constitution. The failure of the Amendment to garner the support necessary for full passage helps to highlight the different strategies deployed by those in favor of passage and Conservative groups that lined up to oppose it.

The amendment flew through the House and Senate in 1972, passing overwhelmingly in both chambers, far surpassing the two-thirds vote needed for passage. It went to the states, where it would need to get three quarters of the states, or 38 state houses, to pass it in order to become a part of the U.S. Constitution. Many states passed it immediately, including, Hawaii, which did so on the very same day the U.S. Senate approved it. But then the pace of states joining the ranks of those that ratified the amendment began to slow. By 1978, the amendment had passed in 35 states, three states shy of the number needed for ultimate passage.

That's where local lobbying became the most intense, carried out by national and local groups.

Organizations lined up to do the work in the states to push, or oppose, state ratification. Groups like NOW that supported the ERA were national organizations with chapters at the state level and a few at the local level that used a range of tactics, from direct action to obtain national exposure for ERA support to lobbying at state capitals. Conservative groups, organized by Phyllis Schlafly's Stop ERA, worked through existing networks—often religious congregations—that functioned like the trans-local networks of the Abolitionists, the Women's Movement of the late 19th Century, and the Civil Rights Movement of the 1950s and 1960s. Lepore has described these members of the nation's evangelical churches as Schlafly's "foot soldiers."[40] These activists engaged in targeted lobbying in key states where they thought they could sway state legislators who would have the fate of the ERA in their hands. While internecine battles within the Democratic Party would lead to defeats of the amendment in several key states, like Florida, the grassroots organizing of Stop ERA would advocate aggressively within state legislators in just enough states to turn the tide against the amendment. Although thirty-five state legislatures would support the Amendment's passage (with five of these later attempting to rescind their support), by the time the deadline passed for ratification in the states (already extended once by Congress), its passage would ultimately fall short by three states.

One of these states was Illinois, where southern, more conservative regions in the state were able to mount stiff opposition to ratification. Political scientist Jane Mansbridge engaged in an in-depth study of the pro- and anti-ERA forces in Illinois to reveal the differences in their organization, makeup, and tactics. The pro-ERA groups in the state consisted mostly of supporters who were deeply dedicated to the cause, but lacked an appreciation for political advocacy at the hyper-local level: that is, with individual legislators on their home turf. As Mansbridge explains it, "in spite of the enthusiasm of its most active members, ERA Illinois was not equipped to organize at the district level."[41] Mansbridge's exposure to the group led her to conclude that ERA Illinois "was an unexciting, frequently disorganized umbrella organization, with few individual members."[42] While it was made up of a number of different groups, "none of these groups had much interest in trying to mobilize its membership to enter active politics in their home districts."[43] More importantly perhaps, the national NOW was gaining members it could bring in through direct mail marketing. This growth occurred as a result of high profile actions like demonstrations and television advertisements that would have less of an impact at the local level, where the battle for ratification was taking place. Such national action helped spur direct-mail membership.[44] Indeed, in the ten years from 1967 to 1978, NOW's membership grew 100-fold, from just over 1,100 members in 14 chapters to over 125,000 in 700 chapters. But NOW's membership could not compare with the size of more traditional membership groups in 1980, including groups like the General Federation of Women's Clubs (600,000) or the United Methodist Women (1,244,000).[45] Even with NOW's boost in membership, linked to its efforts around passage of the ERA, this growth would not ensure state-by-state ratification of the ERA.

The Stop ERA network was different. It was made up of a pre-existing network of local "cells": the congregations of religious groups the members of which were beginning to enlist to oppose the ERA. These groups had certain practical advantages, as Mansbridge points out, like "preexisting meeting places, buses, and claims on their members' time and money."[46] One state legislator interviewed by Mansbridge described the differences in the tactics used by the two groups as follows: "The STOP ERA forces were much better organized and did a better job lobbying."[47] The pro-ERA forces would only appear at the capitol when there was going to be a vote, and then they would have " 'a big hoopla and stage a rally.' "[48] By contrast, the anti-ERA groups were " 'there all the time.' "[49] They were " 'getting to know the legislators developing a relationship with them, not necessarily harassing them, but constantly reminding them of their presence, maybe even talking to them about other issues, establishing a personal relationship.' "[50]

That is not to say that these simple, almost homespun tactics were evidence that the anti-ERA forces were unschooled, undisciplined or unprofessional. One biography of Schlafly revealed her efforts to train and educate the movements' members, including using video technology to record advocates engaged in public speaking on different topics so that they could hone and perfect their advocacy skills.[51] Interestingly, and paradoxically, the anti-ERA forces were so professional that they were able to come across as more grassroots than the pro-ERA forces. Where NOW and groups like it would have their members sign pre-printed postcards to send to their legislators, the anti-ERA forces would have their members handwrite personal messages on their own stationary, and that difference likely sent a message to legislators that the anti-ERA forces were more committed to their cause and might exact retribution from those elected officials at the ballot box should they vote in favor of the amendment.[52]

One cannot overstate the importance of the far easier position in which the anti-ERA forces found themselves. They only had to secure opposing votes from twelve states. In Illinois, for example, a three-fifths vote in the legislature was necessary to secure ratification of the amendment. It was also Phyllis Schlafly's home turf. What they also had going for them was the ability to craft a message that resonated with conservative and even moderate women, women who did not mind some of the "discrimination" women experienced: disparate treatment that, some would feel, benefited women, like being excluded from the military draft and not being assigned to combat positions in the military. While it was unlikely that passage of the ERA would alter arrangements in the military, the anti-ERA forces were able to raise enough objections to the ERA that generated serious questions about its possible impact. In contrast, the pro-ERA forces had few concrete answers to questions about the ERA's possible impact, and were caught in a bit of a bind: minimize the potential political impact and lose potential supporters, or exaggerate the impact and risk being seen as extremist. Complicating matters for the pro-ERA forces even more was the fact that the Supreme Court's interpretation of the Equal Protection Clause of the Constitution had literally shifted under the feet of advocates. The Court ruled that the clause did provide some degree of protection to women as the ratification process in the states was under way. If there was already some constitutional protection for women, why was the ERA necessary? Moreover, one of their key arguments—that the ERA would eliminate disparity in pay between men and women—was hard to reconcile with the fact that the ERA only impacted the action of states and the federal government. It would not change private employment arrangements (and some of the issues related to private employment disparities had already been addressed under civil rights laws). This would muddle the pro-ERA forces' message. They were hard pressed to identify the concrete benefits that women would experience by passage of the ERA, and were resigned to stressing its "symbolic" importance as reflecting a cultural shift more than an improvement of the legal status of women. As Mansbridge explains, the pro-ERA forces had to "keep their arguments vague, since the short-term benefits of the ERA for working women were almost exclusively symbolic, and the long-term benefits were both hypothetical and uncertain."[53]

The anti-ERA forces were thus able to win the message war by hammering a couple of key points, regardless of whether they were accurate: that the role of women in the military would change, that the ERA would eliminate single sex education and sports teams, and it would force women to use unisex bathrooms. These points raised fears among some women; certain types of disparate treatment were welcome. Valuable protections might be lost were the ERA to pass. The anti-ERA forces were able to raise concerns that major—and perhaps unwelcome—cultural shifts might come about as a result of passage of the ERA, while the pro-ERA forces were caught between arguing such shifts were necessary, or were unlikely. Either approach for the pro-ERA forces spelled trouble: arguing for radical changes would lose supporters; pressing for mere incremental change wouldn't rally them. The more radical the change, however, the less likely the cause would attract the moderates that any movement needs to succeed.

Political scientists Joyce Gelb and Marian Lief Palley have studied the impact of the Women's Movement that started in the 1960s on public policy. As they point out, the pro-ERA forces were resigned to arguing for "role change" between men and women, as opposed to "role equity": painted into this corner, Gelb and Palley argue, those forces were unlikely to succeed.[54] Role change arguments are "fraught with greater political pitfalls," they argue, "including perceived threats to existing values, in turn creating visible and often powerful opposition."[55] The pro-ERA forces, in some ways, were forced into this trap. They had to argue that the ERA would be important to rally supporters, without overemphasizing the extent to which it would threaten certain accepted cultural norms. The anti-ERA forces, on the other hand, took advantage of this trap, even if they did not set it. They were able to raise the specter that the ERA would usher in unwelcome change—like single sex bathrooms—that few were ready to accept at the time.

Another lesson from the fight over ratification of the ERA emerges from the assessment of the victory for the Stop ERA forces in Illinois: networked, trans-local organizations that work from the grassroots up are far better at mobilization and agitation than more top-down organizations. These forces were able to leverage personal relationships built in small organizations to further a broader political agenda. They did so effectively and where it mattered: in the state houses and district offices of key state legislators who held the fate of the ERA in their hands. Surprisingly, perhaps, this lesson was not one that resonated on either the Right or the Left, in the offices of national organizations in Washington, DC; New York; or San Francisco. Rather, the trend toward professionalization and centralization, with a movement away from grassroots, trans-local, federated networks, would continue apace, throughout the end of the 1970s, and for the next thirty years.

The final defeat of the ERA would not come until 1982, when the deadline for ratification came, but the seeds of the conservative revival were planted in the 1976 primary fight between Ronald Reagan and Gerald Ford for the Republican nomination for president. The conservative wing of the Republican Party had been mostly marginalized since the defeat of Barry Goldwater in 1964. The dominant wing of the party included people like Nelson Rockefeller, Ford's Vice President, who was recognized for his liberal positions on civil rights, abortion, and government spending. Despite these moderate voices within the Republican Party, a conservative "revolution" would sweep Reagan into the White House in the election of 1980, a conservative revival that still reverberates today. To what extent was this revival a product of the technology that changed the shape of social movements, one that weakened traditional, trans-local bonds, and supported a narrower politics, one based more on interest groups with a more limited focus?

Towards the end of the 1960s, the technology that helped groups form, coordinate, communicate and cooperate changed, and this had profound impacts on social change in the United States for nearly half a century. The ability to engage in mass, targeted mailings would transform civic activism for two generations. Groups such as the National Organization for Women and the Natural Resources Defense Council started in this era, and mostly function to this day as organizations directed from a central headquarters, drawing financial support from what researchers John McCarthy and Mayer Zald would call a "paper membership":[56] dispersed individuals that typically do not really operate in local, grassroots chapters and support the group through making occasional donations (or refusing to donate).[57] As Joel Handler explains, these organizations require little of their members, communicating with most by mail, with leadership claiming to speak for a large constituency based on the membership base but such members may be "only moderately interested in the cause."[58] In contrast, in the early days of the republic, groups operated as local networks that often mimicked both the postal system and the federal government. In the Progressive Era and after, groups maintained the trans-local model facilitated by new means of communication. In the post-1960s, however, national, "bodiless heads," (to borrow a phrase from Margaret Weir and Marshall Ganz) have predominated: groups that function mostly at a national headquarters with minimal connection to or input from their membership base, if they even have one.[59]

Writing in 2003, before the advent of social media, Theda Skocpol described this phenomenon as follows: "A new civic America has thus taken shape since the 1960s, as professionally managed advocacy groups and institutions have moved to the fore, while representatively governed, nation-spanning voluntary membership federations—especially those with popular or cross-class memberships—have lost clout in national public affairs and faded from the everyday lives of most Americans."[60]

Of course, this golden age of civic engagement in post-WWII America was not always so kind to those who were not white, male, Christian and straight. As in the Progressive Era, some groups formed along racial and ethnic lines and thus found it difficult to cross boundaries of social inequality. For example, some unions resisted integrating their roles and all too often opposed integration in the workplace as well. This phenomenon has played out throughout U.S. history. It is no surprise that how a group organizes itself—along racial, ethnic, gender and class lines—often determines the cause it supports. Throughout U.S. history, trans-local groups appear to have had greater success in achieving social change based on the extent to which they would also build bridges along other socio-economic lines. Whether they were cross-class, cross-ethnicity, cross-race, and cross-gender seems to have also had some effect on their ultimate success, particularly as they addressed issues that might have crossed some of the socio-economic divisions.

This forty-year period also saw two different phenomenon unfold: one socio-economic and one social. There was both a dramatic increase in economic inequality as well as a decrease in generalized trust. Is there a link between a lack of trust and broader income inequality? As we have seen rising income inequality in the U.S., we have also seen a decrease in the level of generalized trust.[61] Is the unfolding of these phenomena at the same time mere coincidence? There are likely many reasons for the loss of trust in society and institutions, from Watergate and the war in Vietnam, to the financial crisis of 2008, and there are likely many reasons for the growth in inequality.[62] What we know though, is that at the time that these phenomena were unfolding, there was also a movement away from trans-local organizing; a diminution in robust civic life brought about by such organizing; and few cross-class, trans-local organizations that could serve as a check on such rising inequality.

Returning to that keen observer of civic life in America, Tocqueville believed the cross-class nature of civil society at the time of his visits to the former colonies may have helped promote an ethic of equality, at least among the white community, as Tocqueville observed in one of the opening lines of his sociological study of the early United states: "Among the new objects that attracted my attention during my stay in the United States, none struck my eye more vividly than the equality of conditions." He continued that he "discovered without difficulty the enormous influence that this primary fact exerts on the course of society; it gives a certain direction to public spirit, a certain turn to the laws, new maxims to those who govern, and particular habits to the governed."[63] For Tocqueville, these "habits of the governed" appear to have spilled over into the political realm as well, weakening the power of economic elites in politics.

Central to the Tocqueville's "equality of conditions" as a force in American culture was the connection between this equality and the civic associations he observed throughout the nation. "Thus," he wrote, "the most democratic country on earth is found to be, above all, the one where men in our day have most perfected the art of pursuing the object of their common desires in common and have applied this new science to the most objects." He then asked: "Does this result from an accident or could it be that there in fact exists a necessary relation between associations and equality?"[64]

Tocqueville's observation—that there may be a connection between associational life and equality—is perhaps one of the keenest, and most overlooked, observations of life in the United States that this French aristocrat made during his visits. What was it about associational life that might have helped foster the equality of conditions Tocqueville observed at the time? If Tocqueville's observations about civic life and its impact on economic inequality hold true, the fact that associational life during the period between the Colonial Era and the Civil

War was often cross class, but still monolithic in terms of gender and race, is it any wonder, then, that this period would not see great advances in equality between genders or races? Does the degree of integration of associational life shape the nature of inequality at the time? A nation with robust associations, but those associations are homogeneous, may find that stratification replicated in society as a whole and efforts to combat such stratification unsuccessful.

Is there a connection between trust, social movements, technology, and inequality? At time of great civic engagement—i.e., the post-WWII years, when the GI Bill increased the human capital of millions of Americans—income inequality was at its lowest point in eighty years. And when technology appears to have transformed the shape of civic engagement, making it more centralized and less diffuse, more top down and less bottom up, and more targeted and less cross-class, economic inequality appears to have risen. At the same time, the existence of robust, cross-class groups during the thirty years following the end of WWII appears to correspond to a period of lowered economic inequality.[65] Cross-class and some cross-racial groups flourished, and the Civil Rights Movement saw some of its most significant achievements towards the end of this period. Does the shape of civic activism in social innovation moments, which is often a reflection of the communications technology that enables it to take root and flourish, have an impact on social inequality generally, and economic inequality in particular? I will return to these questions in Chapter 9.

We stand at a time in history when the ability to communicate and collaborate has never been stronger. Yet many feel disconnected from their neighbors and the communities in which they live and work, and less trusting of others and of the institutions that are supposed to serve them. For roughly forty years following the introduction of the computerized mailing list, advocacy groups took a particular approach and form in their efforts to bring about social change. Such efforts rarely included the creation or nurturing of networks of grassroots, trans-local organizations. Nevertheless, as new technology is fostering new ways of communicating, effective groups are starting to use technology not just to facilitate communication. What such groups are learning to do, like so many movements before them, is to build networks and spread messages along those networks that will have a broad appeal. Those messages with broad appeal can, in turn, help to reconnect individuals to networks of change.

The remainder of this book highlights contemporary efforts to advance social change, many of which appear to embrace the traditional methodologies and tactics made popular by social movements prior to the introduction of the computerized mailing list. They are borrowing the components of the Social Change Matrix and adapting them to the 21st century. In the following chapters, I will first explore the potential ways in which new communications technologies can assist leaders in achieving social change by leveraging components of the Social Change Matrix, adapt them to new technologies, and harness a new social innovation moment. I will provide several case studies of initiatives where movement leaders and rank-and-file members of grassroots groups have deployed the latest communications technologies to promote face-to-face organizing, embrace inclusive messages, and brought about positive social change.

5

DIGITAL ORGANIZING

Poignantly, the dramatic strike of public school teachers in West Virginia in the late winter of 2018 may have started because of technology, but not the type of technology I have generally discussed up to this point in the book. While teachers in the state saw their pay, some of the lowest in the country, steadily eaten away by rising health-care costs, their health-care provider was now adding insult to injury. All teachers would have to wear a personal activity tracker that would measure the number of steps each took throughout the day. If a teacher routinely failed to meet a preset step goal, he or she would face a financial penalty, an increase in his or her deductible, and a possible surcharge the insurance company would tack on to his or her premium. For some teachers at least, this was the last straw.

Teachers in West Virginia had slowly seen their insurance premiums, co-pays, and deductibles rise over the years. When Jay O'Neal, a teacher with a master's degree who had been teaching for seven years in places such as San Francisco and Pittsburgh, opened his paycheck one day expecting to see the effects of his annual raise kick in, he actually saw his pay had gone down due to rising insurance costs, which completely ate up that raise. Even with the raise, though, O'Neal was making $6,000 a year less in West Virginia than he was when he started teaching in San Francisco seven years earlier. But it was seeing his pay go down when it should have gone up that made O'Neal angry, and it made him "more ready . . . to do something about it."[1]

What O'Neal and other teachers began to do in the fall of 2017 was monitor the activities of a state administrative body that reviewed and made recommendations to the state's governor and legislature about the costs of health insurance that would be passed on to, not just teachers, but all public employees in the state. He began to attend meetings of that body and met other workers interested in the activities of the Public Employees Insurance Agency, or PEIA. One of those other teachers was Emily Comer, a member of a different teachers' union than O'Neal (there are two in the state, as well as a third that represents non-teachers who work in the school system). They began to monitor the activities of the PEIA and created a Facebook group for other workers covered by public insurance in the state who wanted to learn more about the activities of that body.

In their monitoring efforts, O'Neal and Comer learned that there was a legislative committee that held meetings about public insurance that were open to the public. These were usually sleepy affairs, carried out in the open, but under no one's watchful eye. O'Neal and Comer got a group of interested public employees together to attend one of these meetings. Comer's efforts to livestream the meeting on Facebook faced opposition, however, even though livestreaming should have been permitted because it was supposed to be a public meeting. One of the legislators criticized her for doing so and told her to stop recording the session. This verbal attack was also recorded, and the feed and recording of the interaction went viral in West Virginia.

O'Neal and Comer started to use their Facebook group as a place where individuals could learn more about the activities of PEIA and gain insights into other issues affecting public employees' wages, benefits, and working conditions. They hoped to get a thousand people in the Facebook group by January 2018, and they easily met that goal. It was a private group, and at first one had to know it existed to be able to search for it and ask to join. According to O'Neal, the members of the group were "friends and friends of friends." Members would connect with the other social media networks of which they were members to tell others of the group's existence and recruit new public employee members.[2] Interest in the group heightened, and it was turned into a "secret" group in that knowing it existed was not enough—one could no longer search for it. Anyone who wanted to become a part of the group would have to be invited by someone in the group to join it.

As the group began to gain strength, members started communicating about the increases in health insurance costs. But much of the information the workers received from their employer was in "dry insurance language," as O'Neal describes it. Using the Facebook group, the members tried to share information with other members so they could understand what was going on. According to O'Neal: "We just started trying to post good information about [health insurance policies] and help it make sense for people." At the same time, though, workers were starting to receive letters about "Go 365," the program described at the beginning of the chapter that would monitor workers' physical activity. People started to communicate online, asking about the likely impact of the program, including trying to determine whether their health insurance deductibles could increase and whether there would be a premium penalty if they failed to meet certain activity benchmarks.

The recruitment activity started to accelerate as people became more agitated about the steps program and other changes to their insurance coverage. With the group topping out at about a thousand members in January 2018—they had met their previously articulated goal—someone in the group posted a message asking whether it was time to start thinking about a teacher strike. As O'Neal describes it, "She was the first person to kind of just say out loud what a lot of people had been thinking." The Facebook group exploded in activity, and membership soared. In just a month, it went from one thousand people to over twenty-two thousand.[3] O'Neal and Comer had been the administrators of the group, approving every new member. That work soon became overwhelming. They added six more administrators just to keep up with the volume of prospective members seeking to join the group.

One of the reasons for the jump in membership was that the governor of the state, James Justice, at his state of the union address on January 10, 2018, announced that the teachers would receive a 1 percent raise. For Comer, this was a "huge slap in the face." As she tells it, although she is relatively new to teaching, and was not very active in union politics, it was this declaration by the governor that really lit a fire under everyone, even those that were not political. Before this announcement, Comer says she would stay relatively quiet in the lunchroom and not talk politics or discuss controversial subjects. After that address, "politics was brought up every single day at lunch." She noted that teachers "started talking about it in the hallways . . . [and] you started to openly hear teachers at lunch talk about the Facebook group." After that, "it just kind of snowballed."

But what they were talking about was not partisan national politics. The discussions were limited to West Virginia issues. People were very careful in their discussions "to not let national issues, anything about Trump or anything like that, come into the discussion." For Comer, though, this was not that hard: "It didn't really come up that much to begin with." People knew to bring up such topics "would tear us apart and nobody wanted to tear anybody apart in that moment. . . . We all needed to be together."

The teachers rallied, protested, and ultimately went on strike, conducting a statewide walkout. They used Facebook to coordinate actions. Individuals made signs, created images like one would see in an Internet "meme," and circulated them to the other members of the group. These messages would get shared, and the popular ones would get printed out and serve as signs that individuals would hold up at rallies and in-person gatherings. And that really was emblematic of how the technology worked hand in hand with the street-level organizing. The relationships that were created online would be cemented in person, in face-to-face encounters. As O'Neal describes it, social media would help create relationships, but personal relationships could also jump to the online space. He explains that he would sometimes meet someone at the capitol and the person would recognize O'Neal from Facebook. Other times, they would meet and converse in person and then connect on social media. Face-to-face and online encounters, according to O'Neal, "kind of fed off each other." Some of the things the Facebook group coordinated were small and large actions throughout the state. When information emerged that some industry lobbyists were going to meet with state officials, a protest was quickly organized around the event. When negotiations occurred between the governor and union leaders, rank-and-file members were waiting outside at the capitol. In a stunning development, the governor and union leaders reached an agreement that the members would need to ratify, and news of the agreement reached the protestors outside, thanks, in part, to social media. The individual members were outraged over the terms of the agreement, feeling that the union leadership had not really accomplished much to address worker concerns on health benefits. In the fifteen minutes it took for the union leaders to come out of their meeting and address those gathered on the capitol steps, the rank-and-file members had already galvanized their opposition to the agreement, and those leaders were met with boos.

As word spread throughout the state, and opposition to the agreement gathered steam, pockets of teachers voted in their local communities on whether to accept the deal. Social media proved essential to those who opposed the agreement in terms of ensuring the network would vote down the deal. County by county, the teachers opposed the agreement, and news of such negative reactions to the deal built solidarity and maintained the resolve of others who wanted to stay on strike. Comer said that without the information sharing about the opposition that was rippling throughout the state, "I'm not sure that we would have stayed out [on strike] like we did." The teachers continued their strike until their demands were ultimately met, securing an assessment of insurance benefits that included workers as part of the group that conducted that review.

The successes of the West Virginia teachers spread to other states. Teachers in Oklahoma, Arizona, Kansas, Florida, and other states used the work of the West Virginia strikers as inspiration, sparking widespread uprisings throughout the country. Just as their original advocacy was activated by person-to-person advocacy, fueled by social media, activists in West Virginia are communicating with teachers from other communities, both over Facebook and wherever, as O'Neal puts it, "somebody knows somebody."

O'Neal and Comer are excited about their successes, what they have accomplished, and what they hope others will as well. What lessons do they think they can learn from their efforts? As Comer describes it: "We need organizing, not advocacy." Advocates should no longer just go "to the capitol and kind of ask nicely." For O'Neal, people should not just "accept things the way they are because . . . they don't imagine that it could be better. They don't realize kind of how messed up things are."

Still, despite their success in organizing online, such organizing has "limitations," as O'Neal says. Social media is "really good to get people angry. . . . And I think you can get people motivated to a certain extent." He also thinks "the face-to-face is when they'll go beyond that and do more actions and things like that." For Comer, it's the face-to-face interactions that build the trust from which advocacy can flow: "People trust people that they know in real life." Any organizing activities, and coordination of efforts, "means a lot more from someone you know . . . if it's coming from someone in real life." O'Neal echoed these sentiments. Face-to-face relationships are important because "you can be a little keyboard warrior and do that stuff, but [that] doesn't really make change." It is important "to connect with people face-to-face and get to know them and see each other. . . . I think it made the movement a lot stronger that way."

O'Neal describes their efforts as giving people information, and making it "personal." They gave personal examples of how the changes to the insurance programs were going to affect real people. Having a lot of people working on the information-sharing aspects of their work helped make it a success; they showed the value of working together. According to O'Neal: "The group helped kind of paint a picture of how it could be," of the change that was possible in the world.

When civic life in the U.S. was robust, cross-class networks of translocal civic associations thrived. The times when such associations thrived seemed to be times of greater societal trust and greater social equality (for some). At least in the last fifty years, when these concepts have been measured with greater particularity and rigor, there seems to be an apparent connection between when broad-based civic associations flourished and societal trust was high. This phenomenon is likely a result of what happens when people engage with each other in face-to-face interactions. When those interactions are embedded in a network, that trust can be leveraged and activated to bring about social change. Trust plays a critical role in addressing collective action problems. To the extent we can measure the concept of trust, we also know that generalized trust has been in decline in the U.S. since the early 1970s.[4] In addition, according to Putnam, this decline in trust has corresponded to a decline in social capital.[5] Putnam argues that several factors have led to this decline, including the following. First, an increase of two-earner families means that these two earners have less time to devote to civic activities. Second, those earners have an increase in their daily commute to work, again leaving less time for civic engagement. Third, there has been an increase in the use of "electronic entertainment," especially the television. Finally, there has been what Putnam calls "generational change" as individuals in the so-called Greatest Generation participate less in civic activities and newer generations may be less inclined to do so.[6]

If we are to accept this assessment of the decline of social capital and its causes, does it raise questions for the ability of Americans to come together to solve collective action problems and form social movements? Before we attempt to answer that question, some recent trends may undermine the assessment that Americans are in less of a position to engage in civic activities. First, while the increase in two-earner families is real, for those workers, commute times may be decreasing. With a move back to cities, where many of the jobs are, and technology aiding in telecommuting, commute times are decreasing on the whole.[7]

Second, social media and streaming services offer a dizzying array of entertainment options, as well as opportunities to engage with other consumers of the same media. Indeed, Bruce Springsteen's lament from the early 1990s that there were fifty-seven channels on his television set and yet nothing was "on" now seems quaint.[8] Although the television was in many ways "one-directional," new media offers a dizzying array of options to not just engage, but also connect with like-minded people in a broad network of loose and distant ties. Returning to the media theorist McLuhan, he classified media into two types: hot and cold.[9] The radio is what he considered hot media because people receive the signals this medium sends passively and generally do not engage with the sender or the message he or she sends. Contrast that with what he called cold media, where people engage with a medium and contribute to the force and effect of its message. Unlike radio, new media networks offer users the opportunity to engage with other users, by commenting on, sharing, "liking," and incorporating messages and memes in their own communications.[10] Although McLuhan considered television a cold media, Tom Standage considers it the "opposite of social media" because through it people do not "create, distribute, share, and rework information and exchange it with each other."[11]

Third, the new generational changes in the U.S. mean more and more of the population includes those who have grown up as "digital natives" who are quite fluent in digital communication.[12] As the students from Marjory Stoneman Douglas High School showed in the wake of the shooting tragedy there in February 2018, young people are more adept at using social media than older generations, and seem capable of harnessing that medium to great effect.[13]

Each advance in communications technology in the course of U.S. history has brought with it new capacities for communication, and these capacities have had profound effects on the ability of Americans to pursue social change. This chapter explores the capacities new technologies—specifically the Internet and mobile technologies—offer social movements as they attempt to build trust, foster the growth of social capital, and solve collective action problems. I also explore the ways in which these capacities may strengthen the ability of social movements to utilize the components of the social change matrix—medium, network, and message—to bring about social change. I suggest here that such capacities strengthen opportunities for social movements to make effective use of the components of this matrix. These capacities do not ensure the success of progressive social change. Progressive groups must seek to understand the power available to them to harness these technologies to advance the change they seek, but also recognize the threat they pose to the cause of greater social justice. What follows is an exploration of these new technologies and the ways in which they might help activate some of the components and capacities of the social change matrix described in previous chapters.

New digital networks can be described through the definitional tools of network theory, though digital networks look somewhat different from traditional, analog ones, described in chapter 2. Using some of the theoretical tools we have already deployed, like network theory and social capital theory, the following discussion explores the characteristics of digital social networks and the potential impact of such social networks on organizations, organizing, and social change.

First, digital social networks often flow from the concept of homophily: like attracts like. Whether one is looking for Harry Potter fan fiction, or wants to follow the Australian Rules football club the Darwin Buffaloes of the Northern Territory Football League, social media and the Internet help match individuals to those with similar interests. While this certainly allows like to find like, it can also create information silos: channels through which individuals obtain information only from those similar to themselves.[14] Such filters can keep out conflicting or different viewpoints, limiting our ability to see different perspectives, change our own, and see ways in which our interests may converge with unlikely allies.[15]

While homophily appears a likely factor in the development of digital ties, propinquity (geographic proximity) is not.[16] One of the reasons that propinquity has been a traditional feature of network theory in the past is because the costs associated with communicating are lower when geographic distance is lower. In digital social networks, costs are not associated with physical proximity; rather, the costs associated with communicating across a digital social network are practically zero.[17] This issue, the cost of communication, has been, as the sociologist Paul Starr has shown, a critical component of the creation and adoption of communications technologies throughout U.S. history.[18] With the Internet though, as the legal scholar Yochai Benkler points out, costs associated with the "production and distribution of information, culture, and knowledge" are decentralized.[19] With the Internet and social media, then, one can grow one's own network and connect to the networks of others with relative ease and little cost, increasing one's reach and the network effects to be gained by increasing the size of one's network. While they may be easier to form and expand, it is important to ask whether they also generate the types of benefits that social networks seem to produce in terms of trust and the capacity for collective action problems.

While general use of the Internet began roughly twenty-five years ago, the introduction of the Apple iPhone in 2007 helped launch the mobile revolution.[20] At the same time, we still do not have robust longitudinal studies on the long-term impact of digital networks on trust. Moreover, we have not had good methods for measuring trust and social capital, which has been a problem for social capital research over the years.[21] Nevertheless, some researchers have attempted to gauge the impact of the Internet and mobile technologies on some of the factors that affect our ability to solve collective action problems. As this still-tentative research suggests, digital communications tools, in summary, help build weak ties, decentralize communication and engagement,[22] help communicate norms (both good and bad), strengthen the ability of the members of a network to coordinate their efforts, offer new modes of engagement, increase and amplify network effects, and facilitate effective crowdsourcing. In theory at least, one can see that these digital tools, and the capacities they create, likely can enhance our ability to create social capital and solve collective action problems. At least that is what a theoretical understanding of these concepts would suggest and what we have learned in previous chapters about not just social capital but also the components of effective social change. I will explore each of these capacities in relation to the components of the social change matrix to assess whether we can move from the theoretical to the practical and develop an operational understanding of how to put digital tools to use to facilitate community organizing, strengthen social capital, and bring about social change. I will further test these assumptions in the case studies in the next three chapters.

Medium: Digital Networks, Trust, and the Costs of Cooperation

The West Virginia teachers found social media an efficient tool for organizing rallies, protests, and other activities throughout the state. One of the greatest strengths of digital tools seems to be their capacity to assist their users to coordinate action. As the sociologists Anabel Quan-Haase and Barry Wellman argue:

"What makes the communication possibilities of the Internet unique is its capability to support many-to-many information exchanges among geographically dispersed people."[23] In a study of a community that had recently received digital networking capacity in Canada, the residents of a housing development were able to utilize new, digital tools to coordinate their mass mobilization against their common landlord. As the researchers found, the virtual networks the residents used were "especially useful in reducing barriers to collective action," and enabled the residents to organize meetings, facilitate public engagement, and coordinate member participation in a range of political actions, all in an efficient, effective, and low-cost way.[24] The effectiveness of the connected community was not lost on the target of their activism, who admitted that "the residents organized their protests with unprecedented and unexpected speed." The power of the organized residents was not lost on town leaders either, who noted that they were so well coordinated and active that they successfully prevented the developer from obtaining a fairly routine permit to expand the development, which the residents opposed. Finally, given the resistance he faced from the residents and the effectiveness of their digitally enhanced organizing, the developer told the researchers he "would never build another wired neighborhood."[25]

By making coordinated, collaborative, and collective action easier to carry out, digital tools can, in turn, help build social capital, which can be a product of such action. The act of working together—prodded by digital tools—has the added benefit of building trust and developing the organizing muscles needed to carry out larger-scale actions and efforts. One study of nonprofit groups' use of social media to engage with their constituents showed that such outreach created what the researchers called a "unifying feedback loop": the communication led to action which helped create and foster the growth of social capital.[26] As the residents of the wired community in Canada showed, and as we will see in the case studies that follow in subsequent chapters, virtual tools deployed by a digital social network can enable a community to coordinate concerted action designed to improve the lot of its members. It is no accident that the rise of digital organizing helped organizers coordinate mass protests across the globe, from Tunisia and Egypt, to the Occupy and Tea Party movements. Research by Zeynep Tufekci, an activist and academic who studied the inner workings of some of these movements on the ground, shows that the social-media-activated excitement that these social movements generated at first often disappeared in the face of strong governmental opposition, mostly because the groups had not had the experience of working together over time to build bonds of trust and develop a network or infrastructure to channel the energy of the group for the long haul.[27] At the same time, we know that social media at least played a role in the initial actions that these groups took.[28]

Once again, digital tools helped the West Virginia teachers coordinate their efforts across the state easily and efficiently. As discussed in chapter 3, a great deal of the literature on social movements examines the role that social transaction costs play in movement building: that is, what kind of effort is expected of an activist measured against the reward she anticipates receiving for her trouble. But what if digital tools reduced those transaction costs considerably such that they made the work of community organizing much easier? Would that upset the cost-benefit analysis and encourage more people to get involved with the causes they support? Digital tools make organizing far easier and less costly to coordinate actions than it used to be. In the 1830s, sending a single-sheet letter a few hundred miles cost between one-quarter to one-third of a nonfarm laborer's daily wage.[29] The financial cost of a telegram was similarly prohibitive, not to mention the trouble one had to go through to locate and visit a telegraph operator to send your telegram, with many foregoing the luxury for all but the most important messages they needed to send.[30] Today, digital communications are nearly effortless and individuals can communicate with others, share tips, coordinate events, and spread news stories and videos with clicks, swipes, or keystrokes.[31] This reduction in the cost structure of "information and cultural production," as Benkler explains, has "substantial effects on how we come to know the world we occupy and the alternative courses of action open to us as individuals and social actors."[32] We use LinkedIn, Facebook, and Google search to learn about potential collaborators; deploy social media sites to organize rallies;[33] and promote voting by peers by circulating a selfie outside a polling station or with an "I voted" sticker.[34] This is what the political scientists Lance Bennett and Alexandra Segerberg have called the "logic of connective action": with digital tools, the entire free rider calculus is disrupted.[35] There is no doubt that community organizing is hard. In my work I often felt as Oscar Wilde did when commenting on his reasons for opposing Socialism: there are just too many night meetings.[36] But if meetings are virtual and communications are easy, perhaps some of those meetings will not seem so onerous.

What is more, the combination of the ability to engage at low cost, and to communicate about that engagement, also at low cost, compounds the benefits of new forms of communication. A group of researchers led by the Internet studies theorist Helen Margetts called this sort of information "social information": i.e., "the knowledge that helps people decide what they are going to do with reference to a wider social group and that, in so doing, has the potential to activate people's social norms." As the West Virginia teachers found, the ability to communicate in real time about their activities encouraged others to cooperate, support the effort, and communicate a norm of activism and engagement. As Margetts and her colleagues found: "By providing real-time information about what other people are doing, social media affects the perceived viability of political mobilizations and hence the potential benefits of joining, thereby altering the incentives of individuals to participate."[37] Building on this idea, the political scientist and game theoretician Michael Suk-Young Chwe calls this information "common knowledge": it tells us what others are doing and communicates to the recipient of such information the norm of participation. But it is not just knowledge about what others are doing. It runs deeper; it is "knowledge of others' knowledge, knowledge of others' knowledge of others' knowledge, and so on."[38] This common knowledge can help spur action that becomes a practice, then a habit, then a norm. And this knowledge is what helps spur the logic of reciprocity.

Another benefit of digital networks and the platforms on which they are built appears to be that they offer advocates new modes of advocacy and civic engagement. The West Virginia teachers found livestreaming and digital communications powerful tools in their efforts to inform and inspire their fellow teachers to support each other in their movement to exact better employment benefits. Social movement theorists often talk about the different "repertoires" available to social movements: the tactics they can use to mobilize supporters and bring about change. These repertoires are often fixed in a particular time and place and only change slowly.[39] As Charles Tilly argued, the urban settings of the mid-eighteenth century set the "collective-action repertoire" of social movements for the next nearly two hundred years; mass demonstrations, electoral campaigns, and strikes became the tools of choice for social movements.[40] Today, however, with social media and other means, individuals can not only communicate to the other members of their networks in new and convenient ways, they can also urge those contacts to take part in new forms of action: digital action that often has as its corollary similar, analog versions of the same types of advocacy. As the sociologists Gerald Davis and Mayer Zald argued, these new, twenty-first century tools "enable new repertoires of contention for movements and organizations."[41] One can send an e-mail to an elected official, sign an online petition, "live tweet" an event to raise its profile, or share videos of simultaneous activities, in real time, to build solidarity. Organizers are also making use of other digital tools that help make grassroots organizing easier, like digital scheduling (using sites like Doodle poll), document sharing (using Dropbox and other platforms), instant and easy polling, crowdfunding, and crowdsourcing of messages and ideas. All these tools of engagement in the organizer's toolbox make collective action easier and offer new ways of making one's voice, and message, heard. The use of such tools, at least according to some, signifies the presence of social capital; in turn, these tools activate and leverage such capital for cooperative, collective ends.[42]

Digital Networks

The West Virginia organizers showed that they could build an expansive, statewide network based on connecting, at least at the outset, friends of friends. One of the most consistent and important findings on the capacities of digital communications tools is that they can facilitate the conversion of what we might call "latent" weak ties into more robust ties that serve as the backbone of collective action.[43] Most individuals have many acquaintances with whom they have come in contact over the years: from our school years, old places of employment, friends of friends. Such contacts, if they are not nurtured, are not likely to serve an individual to his or her benefit, through either bonding or even bridging social capital. These weak ties are so weak that they cannot be activated to help an individual get by or get ahead in any meaningful way. One may have lost track of such acquaintances and, without the Internet or social media, have no way of getting in touch with them even if he or she wanted to do so (short of hiring a private detective, of course). By making contact and communication between weak acquaintances easy, digital networks help formalize otherwise latent, weak ties into actionable relationships that can be leveraged in a way that can help both parties get by and get ahead.[44] One study of college students' social media use showed that digital networks facilitated bridging social capital by deepening looser ties rather than strengthening already strong ties: in other words, it helped with bridging social capital more than bonding social capital.[45] Similarly, the members of a community in Canada that had recently received high-speed Internet access used that capacity to extend their respective networks to reach beyond their immediate circle of friends and neighbors, activating and formalizing otherwise latent weak ties.[46]

But such tools not only facilitate communications in the cold dark of digital communications; no, they can also facilitate in-person communications as well. One study of Internet usage in Blacksburg, Virginia, found that digital tools were effective for "maintaining relations and increasing face-to-face interaction, both of which build bonding and bridging types of social capital in communities."[47] Similarly, a study of the functioning of three national groups in today's media environment found that members of these groups (including those in today's American Legion) who accessed their organization's website for information about a group tended to be more active in the group.[48] What new digital tools are doing is allowing communication to happen in a lot of different ways. Through these tools, we can build our networks beyond those with whom we may come in contact through purely face-to-face encounters; we can also activate those loose ties to bring about such face-to-face encounters as well (more on this in a moment).[49]

The "strength of weak ties," as Mark Granovetter explained, is that they are more important to us in terms of improving our lot in life, for solving such problems as finding a better job. But weak ties are also critical in addressing collective action problems. When a small group of individuals in a tightly knit community faces a community-wide dilemma, the strong bonds among members of that community are often effective enough to facilitate the coordination of a community-wide response. Once the problem goes beyond the capacity of a small community to solve it, when it is, indeed, a true collective action problem, ties across communities, which are more likely to be weak ties, are necessary to coordinate action among those communities. True collective action problems thus require not just bonding social capital, but also bridging social capital that can activate networks of networks, building on the strength of weak ties. By connecting these networks of networks, one does not just increase and amplify network effects; the cooperation that follows by activating these networks of networks is likely to create trusting relationships across these networks, as the experience of working together and practicing the "art of associating," as Tocqueville described it,[50] is likely to lead to a lowering of social distance, increased trust, and more cooperation in the future.[51]

From the beginning of their advocacy, Comer, O'Neal, and their many colleagues communicated, using their social media platform, a norm of engagement and participation. This encouraged and led others to do the same. Similarly, when it came time to vote down the initial agreement, communities throughout the state shared with others their decision to reject it. This galvanized the opposition to the deal and ultimately led to its defeat. This ability of new media platforms to communicate norms of participation and cooperation and trigger the logic of reciprocity is one of the key capacities of new communications tools. Conducting surveys that compared participants' digital civil engagement with more traditional forms of civic engagement, the political scientists Miki Caul Kittilson and Russell Dalton found that digital networks "have many of the same benefits for citizen norms and political involvement as traditional civil society."[52] While they admit that social media platforms cannot "completely substitute for bowling leagues and choral societies," they argue that virtual network activity "is most clearly linked to a participatory style of citizenship," through "participatory norms of engaged citizenship," concluding that such activity is "positively associated with several forms of political engagement."[53]

These trusting relationships can lead individuals to action. Indeed, in a study of voting behavior during the 2010 U.S. congressional election, different subsets of 61 million Facebook users received different messages or no messages at all about the voting behavior of their friends on Facebook. One group received a message urging them to vote and information on the number of Facebook users generally who had indicated they had voted through the site. Another group received information indicating the number of their friends who had acknowledged voting, including images of up to six friends who had done so. A third group received no voting-related messages from Facebook at all. The messages sent to the first two groups resulted in an increase in voter turnout by roughly 2 percent when compared to the third group, with the greatest effect among friends whose close friends (as measured by the frequency of their contacts on Facebook) had voted. The researchers posited that it was likely that these close friends also had face-to-face relationships outside of their online connections.[54] A follow-up study of the 2012 presidential election yielded similar results.[55] Social media tools helped boost voter turnout, but mostly through the communications between close friends. Close friendships, those likely built on face-to-face relationships, but boosted by social media capacities, are a powerful combination to turn information into action. That information can communicate a norm about the behavior that is anticipated and expected. We know that communication can lead to a lowering of social distance. When digital tools enhance the capacity of individuals to communicate with each other, particularly when that communication is about cooperative behavior itself—like the decision to vote—they can help those individuals tap into the trust-enhancing qualities of communication, which can, in turn, facilitate trusting and cooperative action.[56]

Thus a key function of digital tools seems to be to enable individuals to convert latent, weak ties into actionable engines of organizing. The West Virginia teachers were able to create a broad network that transcended politics. Relying on the convergence of interests among the teachers—and regardless of their own personal political persuasions—the teachers were able to create a broad network that spanned smaller collections of individuals from different backgrounds to create a robust and durable network that grew in strength when it added new members to the network, but also multiplied its strength by doing so. Returning to network theory, once one does this, and plugs "structural holes" in a network, one extends one's network. This expanded network is stronger, has greater reach, and generates larger network effects by the sheer virtue of the fact that there are more nodes in the network. Such hybrid networks—of bonding and bridging social capital—extend the reach and power of the network and amplify network effects. In the organizing context, as Bennett and Segerberg argue, "the strategic work of brokering and bridging coalitions between organizations with different standpoints and constituencies becomes central."[57] Such efforts strengthen the network while also increasing and magnifying the network effects that flow from the expanded network.

Message: Trust and the Crowd

Another important achievement of the West Virginia teachers was that the rank-and-file members of the union were able to communicate directly with each other. This not only created a new way for them to share information, it also became a means by which they could circumvent the leadership of the union when it reached an initial—and unsatisfactory—deal with the governor. In the 1960s, civil rights leaders were dependent on local and national television networks to carry images of the atrocities being carried out by law enforcement and vigilante groups. Today, with social media, individuals are in charge of the medium. McLuhan argued that every medium becomes an extension of ourselves.[58] Never before has this view been more true. Digital communication is bottom-up, democratic, and decentralized, rather than top-down and approved by third parties.[59] This is critical for solving collective action problems because, as Ostrom has pointed out, successful cooperative action involves including those who would be affected by the norms and rules that govern a setting taking part in the formulation of those same norms and rules.[60] Digital media decentralizes. This phenomenon is sometimes called "disintermediation." It places the power to set the rules in the hands of anyone who would wield it. This desire to be a part of setting the rules that govern one's behavior has strong gravitational pull and was, as the historian Gordon Wood has shown, one of the central rallying points during the American Revolution.[61] What digital tools do is give everyone the ability to take part in norm- and rule-setting, democratizing the capacity for creativity in the social change space.[62] In turn, using the tools the Internet and mobile technologies place at our disposal, an individual can take part in the norm- and rule-setting process, and as a result, she is more likely to trust the ultimate outcome of that process. Furthermore, taking part in that process with others likely builds trust among those participants, lowering social distance and spurring further cooperation.[63] For individuals engaged in the norm- and rule-making processes, the outcomes of such processes become a better reflection of their self-identities and even help shape them through such processes.[64]

Digital tools helped the West Virginia teachers shape the messages they were communicating collectively by using those digital tools to find the effective, galvanizing messages that would resonate, attract support, and buoy the membership through the organizing effort. By combining several of the capacities that digital tools offer—harnessing weak ties, amplifying network effects, generating norms of cooperation, improving coordination of activities, and decentralizing—activists can tap into another force to foster the creation of norms of cooperation and messages that resonate with the different members of the network, creating interest convergences. The decentralized nature of communications within a network means everyone can participate in this norm- and message-generating process; in turn, the norms and messages that emerge from such a process will be, as Ostrom posited, a reflection of the individuals who shape them, leading to greater adherence to them. Digital tools enable organizers—and everyone can be an organizer—to tap into the power of the crowd to generate the ideas, messages, slogans, talking points, and other forms of communication that will resonate throughout the network, sustain the group effort, and even grow the movement. Such a decentralized process leads to greater trust in and identification with those messages.[65] Indeed, when a network adopts a cause, the message that cause promotes is likely a product of the give-and-take that occurs among the members of the group, as participants in the message-shaping process practice, in Tocqueville's words, the "art of associating."[66] He would describe this process as follows: "When an opinion is represented by an association, it is obliged to take a clearer and more precise form. It counts its partisans and implicates them in a cause. The latter teach themselves to know one another, and their ardor is increased by their number. The association gathers the efforts of divergent minds in cluster and drives them vigorously toward a single goal clearly implicated by it."[67] The community organizer Gordon Whitman echoes this notion: that individual members of an organization should take part in setting strategy and tactics through face-to-face dialogue. For Whitman, when this occurs, "people feel a sense of ownership and responsibility for them."[68]

Clay Shirky says digital tools offer the power to organize without organizations.[69] Turning once again to Ostrom's design principle that rulemaking around cooperation should come from members of the group that intends to cooperate,[70] decentralization empowers those members of the group to communicate directly with one another about their willingness to cooperate, to spark that "first trusting move" described by Axelrod, triggering Kahan's "logic of reciprocity." The power to communicate about the contours of cooperation and the cost of noncooperation rests in the hands of the cooperators themselves as opposed to some distant leader or organizer with whom those members are not in contact. Participation in the process fosters trust in the process and lends a legitimacy to the outcome, as the work of another legal scholar, Tom Tyler, has shown: participation in the very process of rulemaking translates into a greater likelihood of compliance with its terms when they are generated through such a decentralized and participatory process.[71] Such participation in rule and norm setting, when it is broad and deep, means those who helped shape such norms and rules have a stake in the outcome, and, as Nassim Taleb, might say, have "skin in the game."[72] In such situations, the final product of the process reflects the interests and needs of the group as a whole, and the norms that emerge are likely more durable and, as both Ostrom and the legal scholar Charles Sabel suggest, more likely to lead to compliance.[73] Simply put, the prospects of cooperative endeavors are strengthened by a decentralized, participatory process that sets the very terms of cooperation.[74] Digitals tools are, by their very nature, decentralized and decentralizing, enhancing the users' capacity to help set the terms of cooperation. That participation ultimately means cooperation is more likely.

A New Kind of Social Capital

Digital networks appear capable of enabling leaders, organizers, and individuals to cultivate relationships, develop stronger networks, coordinate action, build consensus, and activate their strong and weak ties for collective action. In short, digital tools seem capable of generating many of the same benefits that theorists suggest social capital offers. Because those benefits might be easier to generate, is this new type of social capital, what I will call "synthetic social capital," as effective as traditional social capital for generating and sustaining trust in the service of solving collective action problems? When digital networks seem capable of creating the durable networks that lead to collective and cooperative action, synthetic social capital also seems to generate the capacity to solve collective problems.

Synthetic social capital appears capable of generating many of the same capacities as "real" social capital, at least at the outset. It facilitates participation and communication and can generate trust, lower social distance, and help people overcome the challenges of coordination. While traditional social capital does not reside in an individual, but rather in the relationships between individuals, digital tools can replicate those ties and serve as the pathways on which communication and coordination of efforts take place. Digital tools can help foster trust and trustworthiness and can do so with ease and at a lower cost than traditional means of building social capital. This synthetic social capital is not a substitute for traditional social capital but can generate some of the same benefits and can do so with less trouble. Since social movements must overcome the costs associated with participation, if synthetic social capital can bring down such costs, it can enhance the capacity for movements to coordinate action, build trust, and solve collective action problems together.

In May of 1997, a decade before the introduction of the iPhone, IBM's Deep Blue computer did something no one thought was possible at that point: it defeated chess grandmaster Garry Kasparov.[75] It is not uncommon now in certain chess matches for teams of humans joined by computers, what are called "centaurs" in the field, to compete. It is also not uncommon for such teams to take home the top prizes in such tournaments thanks to their ability to combine human intuition with the technologies of machine learning, big data, and superfast processing speeds.[76] Erik Brynjolfsson and Andrew McAfee posit that racing "with the machines" as opposed to against them or without them is going to prove necessary in the present and very near future.[77] Community organizers may benefit from this notion: that digital tools can enhance their ability to build trust, coordinate action, and bring about social change. To this point, I have tried to show that social capital, even synthetic social capital that pulses through digital networks, is an essential component of civic engagement and social movement efforts to address collective action problems. Communications tools—what I have called medium—are central to this effort. But so are two other components: a functioning, translocal network and an inclusive message that touches on interest convergences. These digital tools do not strictly offer positive opportunities for organizations to organize, communicate, and build national movements directed toward positive and progressive social change. In chapter 9, I will explore some of the downsides of digital organizing. But first I will assess the role that new digital tools can play in spurring social change. It is through these additional case studies that I will also explore how individuals are harnessing these new tools to "run with the machines" and bringing about lasting social change in the contemporary age, even as they achieve the benefits that can be generated through translocal networks and positive, personalizing messages that are both built on the very analog ideas of face-to-face contact, shared values, and shared humanity.

6

AMENDING THE VIOLENCE AGAINST WOMEN ACT

In September 1994, after years of grassroots advocacy, Congress passed the Violence Against Women Act (VAWA) and President Clinton signed it into law. VAWA provides federal funding for enhanced law enforcement, social services, and legal services for victims of domestic violence. It also contains special mechanisms that permit undocumented victims of domestic violence to obtain visas and has special rules regarding the treatment of victims on lands governed by Native American courts. The legislation was originally enacted with a sunset provision that required that Congress reauthorize it after five years, a feature that was repeated in subsequent reauthorization bills. In this cyclical reauthorization process, Congress modified and enhanced the law, including adding funding for victims of dating violence, sexual assault, and stalking. Passing renewal legislation became a fairly routine affair, and Congress regularly approved it with bipartisan support. When the law was due for reauthorization in 2011, however, the process did not go as smoothly as it had before. Instead, although VAWA had bipartisan support, advocates wanted to strengthen it with provisions they saw as essential to keeping all survivors of domestic violence safe. These additional protections were met with resistance in Congress.[1]

In advance of the 2012 presidential election, advocates sought to amend the legislation in ways that would strengthen the protections offered to Native Americans, undocumented immigrants, and the lesbian, gay, bisexual, transgender, and questioning (LGBTQ) communities. As this book goes to print, the legislation is up for renewal once again. But in the last reauthorization battle, the ability of advocates to harness the medium, their networks, and their message provides a contemporary example of the social change matrix at work in the age of social media.

VAWA certainly has its critics, from both the Left and the Right.[2] For some, VAWA's explicit emphasis on women in its title appeared to overlook the fact that men in heterosexual relationships can be victims of domestic violence, even though much of the language of the legislation has been modified to make it gender neutral. Critics have also pointed out that the term "domestic violence," especially when coupled with the term "violence against women," evokes old-fashioned images that women are the victims of abuse at the hands of men and that heterosexual relationships are the only ones in which abuse occurs. Such images distort the fact that abuse also occurs in same-sex couples.

As Sharon Stapel, the former executive director of the Anti-Violence Project, describes it, a number of problems arise from using a term like "domestic violence" instead of more inclusive language, like "intimate partner violence." The Anti-Violence Project (AVP) works to fight violence against and within LGBTQ communities. As Stapel explains about AVP's mission, it works "with bi- and gay men as well as trans people, some of whom we would identify as women and some of whom we would not identify as women but were not necessarily included in the definition of violence against women as the legislation was written."[3] The way the VAWA legislation was drafted also precluded LGBTQ victims from benefiting from the programs available through VAWA because the Office on Violence Against Women, the federal agency responsible for overseeing the funding available through the act, interpreted it to say that if a program did not serve at least 50 percent women, it could not receive VAWA funding. As a result, Stapel notes, "programs that were serving primarily gay men were not eligible for funding."[4]

This apparent limitation on VAWA, at least as it was interpreted, had two major impacts on the LGBTQ community. First, service providers did not have funding to serve lesbians, bisexual men, transgendered people, or straight men, and thus had to turn people away who did not fit into this funding structure. Second, Stapel and other LGBTQ rights advocates observed that programs that served the LGBTQ community were not growing at the same rate as non-LGBTQ programs because LGBTQ-serving programs were not eligible for funding under VAWA. In the early 2000s, the sense was that the Bush administration denied funding for LGBTQ programs. According to Stapel, members of the LGBTQ community made reforming VAWA a priority because "we wanted to make sure the language was explicit to include people who don't necessarily identify as women but who still identify as survivors of violence."[5]

Advocates from other communities also highlighted some of VAWA's limitations. For Native American survivors of intimate partner violence, a gap in the law's protections made it difficult to prosecute non-Native American perpetrators of intimate partner violence that took place on tribal lands, where tribal courts have jurisdiction.[6] Critics also claimed that the act's protections for undocumented immigrant survivors of intimate partner violence were limited. Under the act, undocumented immigrants who came forward to report abuse to the authorities could pursue a special visa that would allow them to stay in the United States. This mechanism strengthened the ability of government officials to prosecute those who abused their partners. However, undocumented immigrants who survive intimate partner violence may not wish to report the abuse for fear that the survivor's own immigration status will come to light through the prosecution of the crime and he or she might face deportation. Advocates claimed that the act did not offer enough special visas for such survivors. In addition, although the VAWA petition process would permit a survivor of intimate partner violence to seek a visa for his or her immigrant children, if a child reached the age of twenty-one while the application was still pending, that child could not come under his or her parent's VAWA petition.

As the deadline for VAWA's sunset provision approached in 2011, advocates from a range of communities impacted by intimate partner violence began soliciting input from their constituents and other advocates about the experiences individuals and families were having with VAWA. This effort was an attempt to develop an agenda around reauthorization that could strengthen areas of VAWA they believed needed shoring up.

Rosie Hidalgo, who directs public policy efforts at Casa de Esperanza, has worked to address intimate partner violence for over twenty years, first as an attorney providing direct representation to survivors and later as a policy advocate. For her, the protections available in VAWA to immigrant survivors—even if they are undocumented—are "critical for bringing victims out of the shadows and holding perpetrators accountable."[7] Hidalgo, who was involved in this national listening process from the beginning, described how advocates carved out twenty-five issue areas that touched on different aspects of VAWA of concern to the advocacy community. Those involved in the process formed committees to hold listening sessions to find out where VAWA was being effective and where there were gaps in coverage.

The network of organizations that represents communities across the country—both different geographic communities and different identity groups—realized early on that they needed to work together to pursue the types of improvements to VAWA that each thought would benefit their constituents. In order to do that, they first needed to learn to work together better. "Trust," in Hidalgo's words, "was essential."

Pat Reuss is a former junior high school teacher who went through several VAWA reauthorization efforts as an advocate working for the National Organization for Women and came out of retirement to work on the 2011 reauthorization campaign. She felt that some groups had become insular during the presidency of George W. Bush and were more concerned with maintaining their funding than they were with making sure that VAWA met the needs of marginalized communities. Over the years, some groups made political calculations that led them to defer strengthening the protections in VAWA in favor of securing reauthorization. For Reuss, however, in order to build a strong coalition that would hold together, those demands could be deferred no longer. There was a growing recognition that the network needed to be larger to sustain the campaign to come. As Reuss recounts: "I had promised the [LGBTQ] community that somewhere down the road we were going to take on the issues of that community. With the Native American community we had achieved some protections in earlier versions, but we had promised all along that we would do more."[8] As advocates from such communities pressed for recognition of their needs in the 2011 VAWA reauthorization campaign, Reuss felt it was time to make good on these prior commitments. "I don't have a reputation for breaking my promises," she said. Stapel agrees; she feels that "it's not okay to say 'we'll come back for you.'" Everyone needed to "be on the same bus."

The group that led the 2011 campaign, often referred to by advocates as simply "the task force," was the National Task Force to End Sexual and Domestic Violence Against Women, an umbrella group with over twenty member organizations. This network was connected to a broader coalition of groups: a VAWA "infrastructure" within each state that itself was a coalition of coalitions. Under VAWA, advocates in each state receive funding to support a membership coalition of social service providers and advocates within that state. This state-based confederation is then loosely connected through another group, the National Network to End Domestic Violence. According to Reuss, because of VAWA's success in providing tools to combat intimate partner violence and the strength of the network the law was partly responsible for creating, "we had fifteen years of VAWA having some currency out there."

Representatives of these different coalitions' members came together in Washington, DC, about two years before VAWA was set to expire in 2011. They formed over twenty issue committees that addressed many of the subject areas of interest to coalition members, including youth concerns, health issues, and violence on college campuses. These committees helped put together a "dream bill" that included all the ideas that had filtered up through the committees. The coalition had to make some compromises, but, according to Hidalgo, the leaders of the different committees worked together to reach agreement on what would stay in and what would come out of the legislative package they would ultimately propose to Congress. Everyone tried to defer to each other's expertise on key issues. They learned to recognize the deeply held concerns of different constituencies and tried to accommodate the needs of those constituencies whenever possible.

It was important to the leadership of groups in the LGBTQ community that they join the conversation and become an integral part of the task force as it forged ahead to discuss potential modifications to VAWA in the next reauthorization round. A coalition of groups serving the LGBTQ community asked to be part of the steering committee of the task force. After that request, according to Stapel, the members of the task force had some internal discussions and raised some concerns that, in the political and economic climate, focusing on the needs of LGBTQ people might not be where the group's leaders needed to "spend their attention or spend their time, or their political will or political power."

The LGBTQ community brought political power and influence to the table, however, particularly with the Obama administration. Stapel notes that including that community was the right thing to do morally: "You can't leave [these] survivors behind." She felt that it was time to focus on LGBTQ issues, even if the task force thought that the 112th Congress was not "going to get it done." Stapel believed that although the makeup of that Congress was such that it might not be receptive to demands about LGBTQ rights, the Obama administration was inclined to support such efforts. Because of that, she felt that the coalition had an opportunity to raise LGBTQ issues even if positive change for that community "didn't happen this time." She and the other LGBTQ advocates "were under no illusion and we were convinced that we were not going to get LBGTQ language into this reauthorization of VAWA." Nevertheless, the LGBTQ community felt that it was time to press for a broader and more inclusive VAWA, if for no other reason than doing so would raise the issue and set the stage for future reform. The LGBTQ groups wanted to "shift the national conversation about what domestic violence and sexual violence looked like." The task force ultimately invited Stapel and another representative of the LGBTQ community, Tara Slavin of the LA Gay & Lesbian Center, to join the task force steering committee and take part in the process of setting the stage for the reauthorization fight to come.

Even with an expanded leadership—or perhaps because of it—the listening process required trust among the members of the task force and it took about six months to build it. Hidalgo says that in order to build a strong working relationship, "we realized that the group needed to develop core principles for how we would try to reach consensus around issues, and if consensus could not be reached, how we would work to address them. There was also a core principle to not have any groups advocate at cross purposes." The task force also built trust among its members by assuring everyone that their constituents would not be left out of negotiations on the bill and would not be jettisoned in favor of making a political bargain that aided some members at the expense of others. At the beginning of the process, the organizations representing the LGBTQ, Native American, and immigrant communities were skeptical and sure that the political calculations of the more mainstream groups would mean that they would be excluded from the conversation about VAWA reauthorization at some point. As Stapel explains, "Whenever you are a part of an underrepresented population, there's always the fear that you will be left behind."

The task force developed a document that laid the groundwork for developing trust among its members. In it they agreed to certain shared core values and principles, including that they would all support those who represented marginalized communities within the task force. For Hidalgo, this meant that "certain marginalized communities that faced disproportionate obstacles" could not be left out on "the periphery." Instead, task force members realized that they would "need to move [LGBTQ communities, Native American communities, and communities of people of color] to the center of our work."

To ensure that groups representing different constituents could work together, the network of groups participating in the campaign engaged in team-building exercises. "We would have practicums over the phone in our conference calls," says Reuss, "and each learned about the needs and interests of the groups representing different constituencies. . . . It sounds so corny but we all practically on the phone piled our hands over each other and said, 'All for one and one for all.'" In order to build relationships within the group, the participants had to lower their guard by opening up to each other, sharing their stories and learning about the struggles other communities faced.

For Stapel, the moment when she realized that the interests of LGBTQ communities would not be left by the wayside in an effort to get legislation passed occurred when the president of the National Organization for Women (NOW) said that her group would not support a bill that did not include Native American women, immigrant women, and the LGBTQ community. Stapel says that this "drew a line in the sand": the expanded and inclusive version of VAWA was what the task force would support. Although she understood that at any moment the LGBTQ community might be left out of any legislative compromise that might emerge, "that was the moment where I thought this deal won't be made without us at least having a chance to make our case." For Stapel, the task force had now solidly committed to the idea of not just LGBTQ inclusion but also "inclusion of the three underrepresented populations." At that point, Stapel began to believe that what ultimately happened with the legislation would be a product of the political makeup of Congress and "not a lack of political will on the part of the task force."

It was also important for the coalition's new members to earn the trust of the group. For Stapel, the primary way the LGBTQ representatives earned trust was by offering to do work and then to actually do it. This allowed them to "engage with as many people as possible in the task force so that we could both get to know them and they could get to know us. We could prove that when we said we were going to do something, that we would do it, and we could also, frankly, keep a finger on the pulse of what the task force was thinking." Stapel volunteered to coordinate the grassroots messaging that turned into media opportunities for the task force, but she made sure she was not the spokesperson for the group in every instance. On task force conference calls, which sometimes occurred as often as four times a week, she tried to describe the impacts of intimate partner violence on the LGBTQ community. She and Slavin "needed to make sure the task force members didn't see us as an add-on issue but saw us as integral to the work that they were doing." Stapel also knew that it was important that the voices of all communities be heard in the discussion.

Stapel estimates that from 2011 to 2013 she spent from twenty to forty hours per week collaborating with the task force and working on the reauthorization effort. This involved going to Washington, DC, showing up at meetings, helping spread the task force's message in the media, motivating the grassroots, and getting stories about intimate partner violence out. "We became an integral part of the task force" in an effort to ensure that the LGBTQ community "wouldn't be forgotten" in the campaign to make VAWA more inclusive. She observed the other underrepresented groups doing the same things: negotiating with the task force, discussing issues with their base, getting stories out. All of this "built trust and expertise and a message that people just committed to that 'It's all of us or none of us.'"

Like Stapel, the other leaders who made up the steering committee of the task force had to balance the work of the larger network with the interests and needs of their constituents. While representing their base, the leaders were constantly "bringing feedback from the grassroots to the steering committee so that that feedback inform[ed] the decision making" at the national level. This information also informed the national network's negotiations with members of Congress and their communications with the White House. For Stapel, although the steering committee drove the work of the task force, each committee member had "an obligation to the steering committee and their obligations to their base." Sometimes there were contradictions between such obligations that had to be worked out in the steering committee as a whole before the task force could take a position on something. This balance between representing a particular constituency within the larger network and working for the good of all members played itself out throughout the campaign. After hearing each other's perspectives and the different needs of the communities represented on the task force, a leader from one community was able to convey the stories of other communities as necessary. Reports of such interactions were always communicated back to the larger network, and leaders from different constituencies could follow up with a particular legislator if it might help persuade him or her to support the advocates' position. Over time, representatives of different groups had the opportunity to lobby together and, according to Reuss, "we knew each other's lines. A lot of it was just magic." Part of the campaign involved decentralizing expertise, such that those who were typically in the lead had to learn to step back, let others have a role, and trust each other. According to Reuss: "We taught our nonexpert advocates how to not be afraid to lobby on this really hard issue, but we also taught our experts how to 'let go,' to let the whole cacophony of supporters we had out there take up the banner and demand people support it." Hidalgo notes that it was critical that no "backdoor meetings" took place. Additionally, if someone from a particular constituency could not attend a particular lobbying meeting and an issue affecting that community came up, advocates were able to point to online resources and other materials and to the expert who could speak to that issue, so that those interested in particular aspects of the dream bill would always have resources at their disposal to answer any questions they had about it. For Reuss, when advocates spoke with legislators and their staff, they would "simplify the information" about the legislation "and always give them a resource to fall back on." Reuss recounts how her experience as a teacher taught her some lessons that worked effectively during this VAWA reauthorization effort: "I spent four years teaching junior high school, and I did everything as if everybody is a seventh grader and it still works: building the trust, reviewing information, simplifying information, and getting ready for the test." These tactics would ultimately come in handy when the reauthorization fight required the task force to rely on the cohesion and social capital it had fostered between its members.

Some of the representatives of the more mainstream groups in the task force made the unique needs of the LGBTQ, Native American, and immigrant communities central to how they did lobbying work, even when they faced resistance from members of Congress to some of the changes to the language of VAWA. Unity in the face of opposition helped the leaders and representatives of the more marginalized groups feel that the task force would not sell them out. According to Hidalgo, this "really sent the message at the National Task Force level that everyone had each other's back." In early meetings on Capitol Hill, staff members from the offices of Republican senators and representatives expressed their reservations about many aspects of the task force's dream bill. Even some Democratic supporters urged the task force to consider a bill without added protections, arguing "it would be a lot easier" to get VAWA passed, says Reuss. They told advocates: "If you just went back to the old VAWA, it would pass 100–0." In response, advocates said: "That's why we need a new VAWA. It's time." In other words, it was time to push the envelope.

Some Republicans saw the expansion of VAWA to protect LGBTQ communities as a stalking horse for a bigger agenda. Staff members of one key Republican legislator told advocates, "We know why you're trying to improve gay rights in VAWA: it's your secret agenda to get equal marriage." In response, Reuss said, "You're absolutely right." When everyone gasped, she retorted, "Don't be silly."

But what really fired up the task force in this particular meeting was the attitude of the congressional staff members toward the advocates. According to Reuss, one staffer said, "You guys are all dreamers. They're all going to come after you. They're going to split you up. You're going to lose all of them; so, just face it right now. They're going to get you fighting with each other." Reuss felt that this staffer "did us a favor; he gave us our game plan." In effect, this staffer laid out what the task force's strategy had to be: stick together. For Reuss, "it took an outsider to point out to us how we could win."

Within the task force network the approach was clear: stay unified, and press for enhancements to VAWA, even if it meant that there was a risk that VAWA would not be reauthorized. While some advocates in the field may have said that their preference was to keep the additions but not at the cost of reauthorization, for Reuss, the task force leadership had agreed that it was "'All for one and one for all,' and we said, 'No, no, don't give up.'"

Throughout the two-year campaign, responses from elected officials and their staff helped fuel the determination of the group. In fact, Reuss believes that it might have been the "main driving force keeping everyone together." "We were 'modiculed,'" she joked: "mocked and ridiculed." This was common: "Everywhere we went were told we were asking for too much," explains Reuss. "We got our dander up. Everyone said, 'You can't do it,' and we said, 'At least let us try.'" Part of what motivated the task force to continue the fight to expand VAWA "was that we were told we couldn't."

Early in the reauthorization process, immigrant advocates were placed in a difficult position as competing versions of reauthorization legislation began to emerge. While advocates were looking to increase the number of visas a survivor of intimate partner violence could apply for, a Congressional Budget Office report indicated that increasing the number of so-called U visas[9] would increase the administrative costs of the federal agency that processed the applications. In order to pay for this higher cost, the Senate proposed increasing the cost of another visa. Negotiators in the Republican-controlled House of Representatives pushed back on this proposal, saying that requests for new revenue would have to be initiated in that chamber. According to Hidalgo, this issue became a lightning rod; for her, a "critical resource for immigrant survivors was caught up in the crosshairs" of a procedural fight between the House and the Senate, and that was used as an excuse to hold up VAWA. Additionally, as Hidalgo explained, in the House version, new language in the legislation constituted a significant rollback in survivor rights: it required notice to an abuser whenever a survivor sought a U visa and imposed significant penalties for what might be innocent misstatements petitioners made in their applications.[10] For Hidalgo, VAWA reauthorization had always been about "moving forward," not about cutting back rights for survivors. "In good conscience, no one could advise survivors" to apply for U visas with such onerous requirements, Hidalgo says. In the heat of the 2012 election cycle, the reauthorization legislation stalled.

Nevertheless, the 2012 election year offered the advocates an opening to press for the changes they were seeking. Missteps on the campaign trail by several high-profile Republican candidates exposed what progressive advocates described as a "war on women." Several Republican candidates made statements that many found insensitive or ignorant and offensive. For example, Missouri Republican Todd Aikin, who was challenging Senator Claire McCaskill, claimed that a woman who was raped would not become pregnant if she did not want to because women could "shut down" their reproductive system in such circumstances.[11] He argued that those who cited the need for an abortion in the event that a rape caused a pregnancy were being disingenuous because of this apparent physiological power. Other similar statements had Republicans a bit on the ropes in the election cycle.[12]

VAWA lapsed when a bipartisan agreement could not be reached in 2011 and 2012. But after the Republicans lost the presidential election in 2012 and did not gain as many seats in the Senate as they had hoped they would, the toll of the so-called war on women appeared to have been great, especially when many of the candidates who had been singled out for insensitive statements about women lost. For Reuss, this "rare confluence" of events created an opening to renew thc reauthorization campaign. After mostly waiting until the completion of the 2012 election cycle, leaders from the broad network of advocates worked collaboratively to renew their demand for a reformed VAWA that would be more inclusive. One of the strategies advocates used to promote their message was harnessing the power of social media. Using Twitter and Facebook, advocates urged their constituents to share their stories of intimate partner violence and to express their support for a VAWA with strengthened protections for members of the LGBTQ communities, undocumented immigrants, and Native Americans who had survived such violence. They used the Twitter hashtag "#realVAWA" to share personalized stories of survivors of intimate partner violence and educate the public about what strengthened protections under a reformed VAWA could mean to members of communities who were vulnerable under the wording of the act that had expired.

Advocates shared powerful individualized stories that helped humanize and personalize intimate partner violence. In this way, they utilized effective tools of communication to promote personalized messages. Reuss says that local and state groups pulled these stories together because "almost everyone had been, or knows intimately, a friend or a relative who has been the victim of incest, sexual assault, spousal battering." The message that the task force tried to convey was a simple one: intimate partner violence "happens to everybody; it's not just a secret." Throughout this social media campaign, advocates talked together about what they were doing and shared the stories that were being communicated on blogs and social media platforms. They spoke with each other about whether the campaign was being effective and reported when a member of Congress was following an advocate on Twitter.

In addition, the language they used to talk about this form of violence also shows the effective use of message in their advocacy. In order to combat the exclusive term "domestic violence," advocates in the LGBTQ community and their allies promoted the use of the more inclusive term "intimate partner violence." This language is also unifying and promotes greater equality of treatment because it does not seem to privilege one form of relationship abuse or one type of relationship over another.

In an op-ed, Stapel argued that explicit language referring to LGBTQ survivors was necessary in the VAWA reauthorization legislation to ensure that these survivors would have equal access to services and to help change the general population's attitude toward the LGBTQ community. "Using 'violence against women' as shorthand language for domestic and sexual violence entirely omits gay, bisexual and transgender people from the conversation and renders lesbian, bisexual and transgender women less visible," Stapel argued. Such language had a negative impact on the resources that could aid LGBTQ survivors of intimate partner violence, and "many of the interventions we use as a national response to violence are predicated on the idea that (heterosexual, non-transgender) men abuse (heterosexual, non-transgender) women, which makes the models difficult, if not impossible, to use when working with LGBTQ survivors." This situation could change, she argued, "by explicitly including LGBTQ survivors in VAWA—and by shifting the conversation from ending 'violence against women' to ending 'domestic violence, dating violence, sexual assault and stalking' against all survivors."[13]

For Reuss, changing the language of advocacy posed a challenge, even for those, like her, who were supportive of the effort to strengthen and expand

VAWA. "We had a brand: the Violence Against Women Act," she notes. There was a concern among some that reorienting the language would dilute that brand. "Crest doesn't change their brand," Reuss observed. "When they change their brand, they lose." However, the members of the task force had "promised that we would be inclusive," and Reuss assumed a leadership role in bringing all members of the group along. Because of the relationships she had formed over the years and her credibility with the mainstream groups, she could bring disparate organizations across the task force network together around language that was more inclusive and unifying, language that could help hold a diverse coalition together.

The task force used technology to both promote dialogue among the advocates and mobilize them. Reuss notes that the task force "had the new tools of technology and the beginnings of social media so that when we got on the phone nobody had to pay a phone bill and nobody had to fly in." In the early years of VAWA advocacy, Reuss remembers, they only had the fax machine to communicate their message. But in the latest reauthorization effort, they used conference calls, social media, the Internet, and e-mails to promote their message and stay connected. This meant that the task force could easily draft informative text; share, edit, and post it; and assemble experts who were ready to field questions about particular aspects of the campaign. Digital technology proved particularly useful in facilitating effective lobbying. They could also easily mobilize advocates who could respond to questions posed about the needs of particular constituencies. According to Reuss, in the "old days," she would joke (the time before social media), these materials would have been mimeographed. In 2011 and 2012, however, information was constantly being circulated over social media and e-mail and was always available on websites. When anyone wanted to pass along information, all he or she needed to do was send an e-mail-embedded link to a body of pertinent information that was targeted, curated, and designed to sway public opinion.

Even with the technology that helped them stay connected, Reuss explains that the members of the task force still needed to do "the old fashioned stuff" like talking in person and having face-to-face and "eye-to-eye" meetings. Such interactions helped strengthen the trust that was built and cultivated throughout the campaign. Sometimes the technology created a virtual living room in which the members of the task force could view important legislative milestones together. According to Reuss: "When there were votes, we were all over the United States and we'd watch C-SPAN like we were in the gallery. They'd call the conference number and the group would watch the vote together," even though they were scattered throughout the country.

At the end of 2012 and into early 2013, the negotiations entered an intense phase. The legislative process tested the trust the group had developed through these modern and old-fashioned methods. According to several advocates, the Republican leadership in the House attempted to play different advocacy groups against each other in an effort to promote alternative legislation that would not go as far as the advocates wanted. One advocate described what can only be called a real-world prisoner's dilemma: organizations representing different constituencies were offered deals through which they would get what they wanted in exchange for abandoning their allies from within the task force network. For example, Native American advocates were told that if they supported a weakened version of reauthorization legislation, one that might protect their constituents but did not offer protections to undocumented survivors or the LGBTQ community, the House leadership would support them. Similar tactics were tried on different groups. For Reuss, across the task force, the advocates had become invested in the work of the network, and had developed deep personal connections with each other. As a result of the trust the group developed, the advocacy community fought off such strategies.

"We were singled out and negotiated with separately in this weird kind of way," explains Stapel. The leadership of the LGBTQ community was never really offered anything, but instead, it seemed that some on Capitol Hill wanted to be commended for, in Stapel's words, "think[ing] really hard about how to include LGBTQ people even though you ultimately conclude that you can't." This happened more in the first few months of 2013. "At that point," Stapel explains, "we started talking really vocally with as much urgency as we could with the entire task force as well as the other underrepresented groups that this divide-and-conquer strategy was being used." In the end, though, for Stapel, "nobody was willing to be divided or conquered."

Hidalgo explains that even though the political fallout from Republican missteps of the 2012 electoral cycle was significant and the task force might have pressed their advantage in a partisan fashion, members felt it was important that they not be perceived to be acting in a way that would work with just one party. This theme emerged early in the trust-building process among the task force network. "It was really important in our effort that this had to be bipartisan," explains Hidalgo. "The whole 'war on women' meme was challenging, but the legislators needed to trust us as well." For Hidalgo, "outside forces thought it was helpful in electoral politics to exploit the war on women for partisan benefit." The task force's approach would be different. "We wanted to find a solution and wanted to advance VAWA. We weren't trying to score political points to undermine one party and buttress another. We wanted to send a message to legislators that we were really looking for a solution."

Some Republicans supported the effort to strengthen VAWA. Representative Tom Cole, a Republican from Oklahoma, who is partially of Native American heritage, offered support for certain aspects of reauthorization. A Florida Republican circulated a "Dear Colleague" letter supporting a strengthened VAWA and encouraging her fellow Republicans to vote in favor of its reauthorization. Task force representatives continued to meet with Republican legislators and staffers to educate "rather than demonize" them. As Hidalgo put it, intimate partner violence and sexual assault are "not about one party or the other, this is about all of us unifying. That had to be the message."

House Republicans tried to peel off different groups by offering concessions on some aspects of the reauthorization while trying to take other demands off the table. In the face of these efforts, the group stuck together. In the words of Hidalgo, the task force agreed that its members "were not going to accept any rollbacks. The national effort stood together." For Hidalgo, this "really showed the intersection of issues." Groups were able to see how the issues highlighted by other communities affected their communities. "An LGBT advocate said she also worked with undocumented survivors and they wouldn't let them pit us against each other. Immigrant advocates helped lesbian and gay survivors as well as immigrants." For Hidalgo, the unifying message was clear: "Every survivor needed to be protected regardless of where they were coming from." Even when the task force network was told that certain demands would hold up VAWA reauthorization, "we all stood firm. We made this commitment. We are not going to leave anyone behind." In some instances, the advocates were told that certain aspects of the reauthorization legislation could be addressed later, through a separate piece of legislation, but the advocates were concerned that if anything was left out of the legislation, it would be difficult to revive it. The feeling of the task force, as Hidalgo explained, was that "this was the moment."

This approach proved successful in many respects, as the version of the final reauthorization legislation that passed both houses of Congress in a bipartisan fashion contained almost all the protections the advocates sought. Yet there came a time when it became evident that the U visa piece could hold up the reauthorization. Hidalgo feels that the expanded components of VAWA served as a "veil" the House could hide behind when it would not address the U visa issue. "They had a good sound bite," she explained, which gave the opposition some traction: they were expanding VAWA, even if they were not addressing all of the new components the advocates wanted. However, immigrant advocacy groups throughout the country deferred to the immigrant advocates among the task force leaders. When Senate leaders promised the group that they would include the U visa protections in immigration reform legislation that was on Congress's future agenda, the immigrant advocacy leaders felt comfortable dropping the U visa demands in VAWA, mostly because they knew the other members of the team would support the cause of immigrant survivors in the future.

Hidalgo and the other champions of immigrant survivors of intimate partner violence did not want the U visa issue to be an issue that could hold up the legislation. The word from grassroots advocates was that many immigrant women were receiving "deferred action" from immigration officials that allowed them to stay in the country legally and obtain certain benefits, like work authorization. As a result, there was little need to press for the U visa expansion at the time, especially if it jeopardized the other enhancements task force leaders thought they could obtain through the reauthorization process.

Leaders in the task force had to sell this position to the grassroots groups pressing for added protections for undocumented survivors, however. They met with other national organizations to explain the complexities of the law and the political calculations and concessions the group thought were worth making. These leaders stated that a lot of the improvements in the legislation would protect the Latinx and immigrant community. According to Hidalgo, the enhanced protections—for teens, for survivors of sexual assault, for the LGBTQ community, for Native American survivors—would all help the immigrant and Latinx communities too. The message was that the immigrant advocates "had to take a stand in solidarity" and the other Latinx groups "had to trust us." She added that "part of our role was to be a bridge" and let the broader community know that "we were part of the process and helped to shape the advocacy. We were still getting deferred action and work authorization. No one was being harmed."

The state of the eleventh-hour negotiations did not faze Stapel, partly because of the trust that had been developed within the group and partly because of the relatively low expectations the LGBTQ community had set for this round of negotiations on VAWA reauthorization. "Anything would have been progress for us," says Stapel. "Even legislative history"—the record of congressional deliberations about the bill that showed that Congress was thinking about LBGTQ issues when discussing VAWA reauthorization—"would have been progress." Thus much of what the LGBTQ community had set out to accomplish in the reauthorization fight had already come to fruition. "As long as folks were talking about" LGBTQ people, she said, and as long as the task force was "really committed to the inclusion of" the LGBTQ community as fundamental to the task force's work, it "wasn't going to be a shocker to us" if the task force was not able to get inclusion of LGBTQ people in the reauthorization. "Our primary priority was to shift national conversation, not to pass legislation the first time around," Stapel said. As long as the task force members kept talking about LGBTQ people and "as long as we kept getting invited to the table, as long as the *Washington Post* kept doing stories about" LGBTQ survivors, "and as long as [members of Congress] kept getting asked about whether or not they would include" members of the LGBTQ community, "we had created this incredible success based on our initial goals before the legislation even came up for a vote." Stapel was pessimistic about the prospects for a fully broadened VAWA until the very end. "Even a week before VAWA passed I didn't think we were going to get it done." The best that many in the LGBTQ community thought they could accomplish was to "set the stage" for the following reauthorization.

In the end, the Republican-controlled House supported much of what was in the Senate's version of the legislation. Knowing they could obtain much of what was originally in the dream bill, the network of advocates felt they could make a deal that included protections for LGBTQ communities and others. Ultimately, the Republican leadership in the House relented and let the reauthorization law come to a vote. It passed with bipartisan support, by a vote of 286–138 in the House; eighty-seven Republicans joined 199 Democrats who voted to accept the Senate version of the legislation, which included strengthened protections for survivors from the LGBTQ community, undocumented immigrants, and Native Americans.[14]

The successful VAWA reauthorization fight that culminated in 2013 serves as a contemporary version of the social change matrix. It was fueled by modern technology but was still grounded in personal connections, trust, and social capital. The members of the task force used modern technology to communicate among themselves and to encourage a national conversation about intimate partner violence. Internally, task force leaders used simple mechanisms, like conference calls and e-mail, to communicate in real time about strategy and to educate each other about developments in the advocacy campaign. They also used these technologies to build trust. More communication between members meant that everyone could be kept abreast of communications with legislative leaders and their staffers, so that no one felt left out of the advocacy conversation. Because groups representing marginalized communities were afraid that their issues might be dropped in favor of narrower demands of more mainstream groups, this type of communication convinced all members of the task force network that no conversations were taking place that could lead to side deals.

While the enhanced ability to communicate in real time over great distances helped build trust, such trust was also facilitated by old-fashioned in-person meetings and by individual advocates sharing their own stories and those of their constituents. This information sharing helped members, in the words of Pat Reuss, "become vulnerable." In turn, the personalized stories, when shared over social and other media, also helped connect the advocacy to a national effort and a national dialogue.

The inclusive and unifying language that was at the center of the group's messaging—both internally and externally—accompanied the personalized nature of the communications. By moving beyond the term "domestic violence," and embracing the broader term "intimate partner violence," the task force was able to connect to groups that had previously felt excluded from much of the advocacy around these issues. By using more inclusive language, the network was able to both attract the LGBTQ community to the campaign, as well as make a commitment to keeping the issues important to that community at the center of the VAWA advocacy. They were also able to attract, activate, and sustain a truly translocal network, one made up of a national coalition of groups that reached into every geographic community and many other types of communities based on the identities of their members. It was a network of networks that spanned the nation and allowed an intersectional alliance to form, one that was built on trust, was stronger together, and stuck together even when they faced efforts explicitly designed to divide them.

This blending of the old with the new, a combination of the latest means of communication and a diverse advocacy network that promoted a unifying and personalized message, reflects the social change matrix in action in current times. As we will see in the next chapter, this combination is no accident, and is proving that it can be effective as a means of bringing about social change in other contexts as well.

7

MARRIAGE EQUALITY IN MAINE

On November 6, 2012, millions across the world watched the election night returns to determine whether Barack Obama would earn a second term as president. While this contest was decided relatively early that evening, many also watched late into the night, awaiting the results of four state ballot referenda regarding same-sex marriage. In three of them, Maryland, Washington, and Maine, voters were considering whether to approve same-sex marriage. In Minnesota, voters were considering a constitutional amendment to ban same-sex marriage there. Just eight years earlier, in 2004, Republican leaders had advocated using ballot referenda outlawing same-sex marriage as a way to drive conservatives to the polls. Some believe that this helped defeat Democratic challenger John Kerry's bid to unseat President George W. Bush in key states like Ohio and Missouri.

In 2012, marriage equality advocates scored a clean sweep, coming out on the winning end of all four ballot initiatives. This change—in just two national election cycles—may be one of the quickest reversals of political fortunes in U.S. history. At least some of the success of marriage equality advocates may be attributed to their effectiveness in using the principles of the social change matrix in their efforts.

Advocates for same-sex marriage had often pressed their case by arguing that it was unfair to deny gay and lesbian couples the bundle of rights that go with marriage: the right to visit a hospitalized partner, the right to claim insurance benefits, and others. Probably the lowest point in the fight for same-sex marriage occurred in California in 2008, on the night the nation elected its first African American president. Even though President Obama carried California by over twenty percentage points that night, a referendum banning the state from recognizing same-sex marriages—Proposition 8—passed by a 52–48 margin, carrying a majority of African American, Latinx, and Asian voters. This stunning defeat in California in 2008 led same-sex marriage advocates to engage in an intense review of their tactics and message. After some deep soul-searching, they agreed that the message had to evolve.

Advocates decided that they would focus more on the fact that same-sex couples wanted to marry for the same reasons that heterosexual couples did. For marriage equality advocates, this was a stronger message to promote than one that hinged on the protection of legal rights alone. A renewed campaign relied on the rhetoric of equality, fairness, and dignity, not on legalistic arguments. Even the language they used when talking about same-sex marriage would have to change. Instead of using "same-sex marriage" or "gay marriage," advocates, following the early lead of some, began adopt the term "marriage equality," or, as marriage equality advocate Evan Wolfson puts it, "marriage."

In 1983, when Wolfson was still a law student at Harvard Law School, he wrote a paper for one of his classes that outlined a strategy for recognizing marriage for same-sex couples. This paper, consisting of 141 pages with 710 footnotes, could have easily turned into an overblown piece of legal scholarship. This one did not. In it, Wolfson charted a legal strategy centered on equal recognition of same-sex marriage as a moral imperative grounded in the dignity and respect that the U.S. Constitution requires federal and state governments to afford every individual. He wrote: "Human rights illuminate and radiate from the Constitution, shedding light on the central human values of freedom and equality." Wolfson argued that all people, "as individuals, possess a transcendent personality of capacity to choose, to make themselves, and to shape somewhat their lives. This freedom, this autonomy, is our most precious human attribute."[1]

Wolfson argued that the state is not "an equal party" to the marriage agreement; rather, it is merely "an audience."[2] Instead of emphasizing strictly legalistic arguments in his paper, Wolfson embraced a broader vision of marriage equality and of marriage itself. While he made forceful arguments about the legal underpinnings of the right of gays and lesbians to marry their partners, he also made the case for same-sex marriage as a moral imperative that was bound up in what it means to be fully human.

For Wolfson, same sex marriage had to be recognized on both constitutional and moral grounds with "equal respect and joy."[3] It took nearly three decades for a movement to coalesce around framing the issue in not only legal but also philosophical and moral terms, but that refined message is the one marriage equality advocates ultimately embraced. It is also one that the Supreme Court shared when it determined that marriage equality would be the law of the land. In this chapter I offer a brief history of this campaign, focusing on efforts in Maine in 2012, where advocates fought to pass a pro-marriage equality ballot referendum. This state-based campaign is an example of an effective deployment of the components of the social change matrix. The campaign also played a part in the larger effort to promote marriage equality on the national level.

After Wolfson graduated from Harvard Law School in May 1983, he began working at the Brooklyn district attorney's office. In his free time and on weekends, he began volunteering for Lambda Legal, an organization that describes itself as the "nation's first legal organization dedicated to achieving full equality for lesbian and gay people."[4] In 1989, Lambda hired him as its attorney. His twelve-year tenure there included work on a landmark case that legalized same-sex marriage in Hawaii and litigation of a case involving discrimination on the basis of sexual orientation in the Boy Scouts, a case that went all the way to the U.S. Supreme Court.

In 2001, Wolfson began building his own organization, Freedom to Marry, which focused primarily on obtaining marriage equality across the country. Freedom to Marry formally launched in 2003, after Wolfson spent two years getting buy-in, raising financial support from the community, and building on work he was doing through his self-described "ad hoc leadership" on the issue at Lambda.

The defeat at the polls in California in 2008 forced the movement to assess the campaign for same-sex marriage, its strategy and tactics, even the very language that was used to describe the concept. The goal of the review was, in Wolfson's words, to "really find out through a genuinely open process of brainstorming and research to determine what was most likely to work." According to Wolfson, Freedom to Marry analyzed why the movement lost the vote, first on Proposition 8 in California in 2008; and then in Maine, in 2009, in a ballot referendum that overturned legislative recognition of same-sex marriage. Freedom to Marry and its allies set out to determine what they needed to do to reach not only "the fragile majority" they had built by 2010, but also the next 5 to 15 percent of what Wolfson called "the reachable but not yet reached middle." They analyzed more than eighty-five data sets of focus groups, polling from five years of campaigns (including the California campaign), and the experience they had gained over years of advocacy.

The goal of the effort was to analyze how the movement could make the case for marriage equality "in a way that would bring along the next group of Americans and solidify the national majority we were just then on the cusp of attaining." For Wolfson, "that process led to us understanding that part of what we were doing in our mix of messaging and mix of messengers was effective but part of it was also not reaching the very people we needed to reach with what they needed to hear." In the end, the process validated what many, including Wolfson, were thinking at the time.

The assessment concluded that a shift in approach was necessary. The movement needed to emphasize "the universal and common appeal to shared values framed in personal stories and an emotional connection." Instead of talking about "rights and benefits and legal justice and constitutional abstractions, all of which are very important and very legitimate," Wolfson explains, "we would do better by leading with and emphasizing the other legitimate and authentic and real and important parts of our case." For Wolfson, these were the personal stories, such as stories about "how people were affected by the denial of marriage and nongay family members who care about their gay family members" and the "shared values of love, commitment, family, connection." Wolfson believed that it came down to "the Golden Rule": that is, "treating others the way you would want to be treated." In these ways, the movement strived to "emphasize the heart instead of leading with the head, though both are legitimate and important and both contribute to building a majority." The subtle shift in messaging that the advocates would adopt was consistent with what was happening with the rhetoric around same-sex marriage in society and culture more generally at the time.[5]

Consistent with this theme, the campaign's leadership learned at least one positive message from the loss in the Proposition 8 battle. In one community, Santa Barbara, a television ad aired that depicted a bride facing hurdles on her way to the marriage altar. It concluded by asking, "What if you couldn't marry the person you love?" Polling conducted in a similar community, Monterey, during the period when the ad was shown in Santa Barbara, showed that support for same-sex marriage remained unchanged there. In Santa Barbara, however, it rose by eleven points.[6] Data points like this helped chart a course for a different way of framing the fight for marriage equality. Wolfson had always advocated a shift away from a legalistic defense of same-sex marriage to "freedom to marry, or marriage," even before the defeat of Proposition 8. Wolfson says that he "was . . . known for telling people, 'We don't want gay marriage, we want marriage.'" Wolfson argues that although "talking about law and rights and benefits and protection" is "very legitimate and real," at the same time, "it's not about coming up with some kind of sound bite or tagline that would somehow magically change hearts and minds, but it is about emphasizing the explanation and the connection in a way that would reach the people we were trying to reach most effectively, and that's what we got much better at doing."

This "new" message was not, in fact, new, however. Wolfson had written about this kind of approach in his law school days. Others had also talked about the need for marriage equality for the LGBTQ community in similar terms; it's just that the message of legal rights seemed to dominate the discourse within and outside the movement for almost two decades. The journalist Andrew Sullivan had also pointed out that the most persuasive argument in favor of marriage equality should emphasize the LGBTQ community's desire to share in the institution of marriage, but it would reflect shared values in both the straight and gay communities. In 1989, Sullivan wrote an opinion piece in the *New Republic* entitled "Here Comes the Groom: A (Conservative) Case for Gay Marriage."[7] In it, he criticized the growing trend at the time of states recognizing domestic partnerships—an alternative to marriage equality—for both gays and straights. While domestic partnership gave some rights to the LGBTQ community, for Sullivan, it did not go far enough.

For Sullivan, marriage equality "squares the several circles at the heart of the domestic partnership debate." Marriage would allow "for recognition of gay relationships, while casting no aspersions on traditional marriage." In fact, marriage equality "merely asks that gays be allowed to join in." Marriage equality, rather than domestic partnerships, "harnesses to an already established social convention the yearnings for stability and acceptance among a fast-maturing gay community." The recognition of marriage equality would "foster social cohesion, emotional security, and economic prudence." For Sullivan, recognizing marriage equality would not deny so-called family values; it would reaffirm them.

In 1995, Sullivan followed this opinion piece with *Virtually Normal: An Argument about Homosexuality*, in which he continued to stress the ways that arguments about marriage equality should be based in what the heterosexual and LGBTQ communities share. He argued that "it's perfectly possible to combine a celebration of the traditional family with the celebration of a stable homosexual relationship." For Sullivan, "the one, after all, is modeled on the other."[8] Such arguments centered on equality. A political argument based on the equality of the heterosexual and LGBTQ communities is one that could harmonize the legalistic arguments he believed were most often espoused by liberals with those based on the elevated status of marriage that conservatives might make. This status is affirmed by recognizing marriage equality and is not undermined by it.[9]

These different themes percolated for nearly two decades, as individuals like Wolfson and Sullivan advocated for a greater emphasis on the social benefits of giving the LGBTQ community equal access to the institution of marriage. Groups like Freedom to Marry and other advocates, according to Wolfson, were "not asking for anything separate or other or lesser or different." What advocates were seeking, Wolfson added, was for lesbian and gay couples "to participate fully and equally in the same freedom to marry that the constitution respects and that people cherish, and that nongay people know." For Wolfson, talking about the demand for the recognition of marriage for same-sex and heterosexual couples alike in terms other than those based on equality weakened the position: "Any time we modify it or qualify it or make it 'other' we are only distancing ourselves from the very hearts and minds that we need." Wolfson felt that Freedom to Marry should be at the forefront of renewed campaign messaging that would ultimately embrace a "style of engagement and message" that helped "close the chapter on Prop 8" and begin writing a new chapter in 2012.

Freedom to Marry began to reposition the movement to embrace the themes of love, commitment, and values, and to help, in the words of Wolfson, "shift our discourse and then *the* discourse." Although some had been advocating talking in more emotional terms, the organization's review of campaign rhetoric determined that the dialogue was still focused on the legalistic arguments. When the campaign moved to Maine after 2009, it made a concerted effort to emphasize "neighbors and shared values and love and families" instead of the more formal, rights-based arguments used by prior campaigns.

But the effort required more than messaging. It also required fundraising; building an infrastructure that could get the message out; and communicating the message through both media efforts and face-to-face, personal engagement. As Wolfson is fond of saying: "There is no marriage without engagement." After the defeat of the Maine ballot initiative in 2009, the movement set its sights on the state again and began to build the infrastructure for a new referendum on the ballot for the 2012 presidential election cycle.

A central component of the new Maine campaign was canvassers who had one-on-one conversations at front doors in towns and communities across the state. Their goal was to identify those who could be persuaded to support marriage equality at the polls. They did this by engaging residents face-to-face. The commitment to engaging in such conversations was in part a result of some of the analysis of what had gone wrong in California. Part of that effort included a so-called postmortem: a door-knocking listening tour in which canvassers visited sixty thousand residents to find out why people may have voted for or against Proposition 8. While canvassers had gone door-to-door in the buildup to that vote in California, they usually had very brief conversations in which they pitched opposition to Proposition 8. These conversations were scripted and short and there was little time for the canvasser to have an in-depth conversation with potential voters.

Freedom to Marry learned from its assessment of the outcome of the vote that longer conversations might work to persuade voters, conversations that probed people's "core values" and got them to talking about some of the newer themes the campaign wanted to emphasize, like love, commitment, and compassion. These concepts could help people connect with the marriage equality message on a personal level. The new campaign tested a variety of strategies, including door-to-door conversations, mailings, and phone calls. They also tested the messaging itself.

In one field test, the Maine campaign conducted a series of phone calls, some of which involved the use of persuasive techniques in an attempt to convince the target audience to support marriage equality and others that simply used polling to gauge what level of support already existed. Over six weeks, the campaign called twenty thousand households. Twelve thousand households received a call in which the caller tried to persuade the recipient to support marriage equality, and eight thousand served as a control group and were asked only about their level support for marriage equality. An independent group made follow-up calls to determine the effectiveness of the two strategies. Those who had received persuasive communication phone calls were 12 percent more likely to support marriage equality than those who were simply polled. This outcome informed the campaign that persuasive efforts worked. The demographic information about the recipients of the calls also offered insights to the campaign about which types of people could be persuaded to support marriage equality. For starters, they found that men were generally more persuadable than women, although in general women were more supportive of marriage equality.

After the campaign received the results from this field test in May 2011, it spent the following summer developing leadership and training canvassers and organizers. Members of the campaign had videotaped some of the in-person conversations that had already taken place, and they had recordings of some of the phone calls that were made during the field test. The campaign used these recordings to train volunteers and paid staff about how to conduct conversations, how to follow them where they might lead, and how to attempt to direct people to topics that would make them the most comfortable talking about this issue.

Learning from prior experiences and their ongoing assessments, the Maine campaign did door-to-door canvassing in which trained volunteers and campaign staff asked more personal questions of potential voters. If an individual was married, the canvasser would ask why that individual had married and what it meant to him or her to be married. The canvasser might also ask whether the individual knew someone who was gay or lesbian. One of these canvassers was Amy Mello, a veteran field organizer for marriage equality campaigns who had worked on the Proposition 8 campaign in California and on other efforts in Massachusetts, Connecticut, and Rhode Island. For Mello, who estimates that she had up to seven hundred conversations during the Maine campaign, such conversations were geared toward getting voters to "open up," to talk about their core values. This often led to a conversation about the Golden Rule: the importance of treating others the way one would want to be treated.[10]

Instead of the scripted conversations like those that took place in the lead-up to the Proposition 8 vote, these conversations were more loosely structured, allowing the canvasser to follow up as he or she saw fit and gently direct the conversation toward a topic that might resonate with a prospective voter. According to Mello, if the voter was talking about his or her faith, the canvasser would ask four or five questions on the subject that might gently probe ways that a person of faith might be convinced to support marriage equality. If the individual asked about tradition, there would be four or five questions that addressed and explored that topic. For Mello, some of the work Freedom to Marry and other allies did on messaging gave the canvassers "really good themes" that might strike a chord with the prospective voter depending on his or her responses to the dialogue with the canvasser. The campaign was confident that this type of approach was the one that would prove to be the most effective.

According to Mello, the feedback the campaign received allowed it to adapt on the fly. Canvassers sometimes showed prospective voters different video commercials that the campaign planned to air and asked them what they thought about those spots. "We were part of the engine of the communications," Mello explained, and through their efforts, the messages and approach "could be tested in real life." The constant feedback the campaign received consistently showed that through the door-to-door efforts, the canvassers were really addressing the personal concerns of prospective voters and learning which messages worked.

It was labor- and resource-intensive work, however. Such conversations could last for up to twenty minutes. According to Mello, the campaign believed it made sense to invest in "the most expensive, longest conversations" and that success was measured in inches, not yards. On average, a volunteer would have about sixteen conversations a day with prospective voters. About eight of those voters would be undecided or opposed to gay marriage. Mello estimates that on a typical day, a canvasser was likely to persuade just one person to support marriage equality.

The campaign turned these incremental victories into a statewide win through sheer volume. Throughout the life of the campaign, canvassers had 320,000 of these conversations and reached 240,000 voters (sometimes there would be more than one conversation per voter). Through their efforts, they were able to reach roughly 25 percent of the likely electorate.

Mello believes that these personal conversations had a tremendous impact. They created a "buzz" as feedback came back to the campaign that prospective voters—even those who lived deep in the woods of Maine—reported that someone had come to visit them to talk about marriage equality. Nothing like that had ever happened before. In follow-up phone calls, those voters recalled when someone had come to their door and talked to them for fifteen to twenty minutes about values. This was new. It was also effective.

"We saw in a year a dramatic increase in the support for gay marriage," says Mello, and the information the campaign was gathering indicated that this support was unwavering. "We held voters because the connection with them was real," Mello explains. "They made commitments to us. They had a real person in their heads" when they thought about the issue. And the stories that canvassers shared resonated with some voters. If a canvasser went to a home where another member of the campaign had already visited, the resident would report that someone had already been there and would be able to tell the story that the canvasser had shared. It was clear that these conversations were having their intended effect.

Mello feels that the trust the canvassers were able to develop with prospective voters was central to the success of the canvassing effort. They accomplished this by sharing their personal stories and their own reasons for supporting marriage equality. Their willingness to be vulnerable helped convince prospective voters to open up and share their own experiences with marriage, with individuals who were gay or lesbian, or with the issue of marriage equality specifically. "We were sharing as much as we were asking," Mello says, and this had a powerful effect on potential voters.

Such connections did not come instantly. This might help explain why the polling phone calls that the campaign did early in the campaign were not as effective as calls that attempted to persuade voters. These sorts of conversations took time and effort. As Mello explains, in the first thirty seconds of a conversation, the initial statement many canvassers made would be met by a "stock answer" from the prospective voter. About twenty minutes into the conversation, however, after the canvasser had been vulnerable and shared his or her personal connection to the marriage equality issue, the voter would talk about broader concepts, and then he or she might, in Mello's words, "talk about a cousin who died of AIDS." In Mello's experience, this was common. When the campaign began, she was skeptical about whether these sorts of conversations could occur because she had heard that people from Maine were "not the most open, sharing-type people." In the end, though, she was impressed by how the canvassers were often able to connect with potential voters on a personal level.

The canvassing effort required a combination of paid workers and volunteers, but because the work helped canvassers connect to prospective voters on a personal level, the retention rate of workers and volunteers was high. Unlike some other campaigns Mello has worked on over the years, "our canvassers never quit," she said. Some people worked with the campaign for two years: "They were part of the process of learning and making the script better, and they really took it that there was a job to do." The best canvassers conducted over a thousand conversations over the life of the campaign. According to Mello, for many, the canvassing experience was "transformative for the canvasser and transformative for the voter." For Mello, this feeling was personal; through the canvassing effort and having these powerful conversations with so many voters, she "developed a different understanding" of herself. She feels that the "best part of the job" was doing the canvassing. Powerful stories emerged from prospective voters during the canvassing about their connection to the issue of marriage equality. In order to broaden the local and personal appeal of the campaign's message, the campaign worked to incorporate such stories into the statewide marketing effort.

This combination of in-person conversations and message-driven advertising helped seal the victory in 2012. According to Wolfson, an effective campaign must first have the right message, and then it can go on television, which can serve as an "echo chamber" for the message. Such efforts would "elevate the discussion and set a frame for personal conversations" that would follow in the door-to-door work. In other words, the advertising was a complement to the canvassing. Wolfson says that "what closes the deal is the ground game of personal engagement and personal conversations that really ask people to come with us on the journey." For the first time, as support for the cause was growing, the marriage equality campaign had the funding and infrastructure to have both an effective marketing strategy and an effective canvassing effort that reached a large percentage of the electorate. For Wolfson, both air and ground efforts are necessary: "The air cover is an important element of a good campaign but it is not a substitute for making that ground game happen." Both Mello and Wolfson believe that trust played an important role in ensuring that person-to-person conversations were effective. In Wolfson's words, "the air cover alone is not enough." For Wolfson, the ground game is what builds the trust: "The personal engagement, the personal conversations, the personal messengers; not just the messengers but the messages people can connect with in a personal way." In all of this, trust is "absolutely essential."

Results of a survey of voters taken after the electoral victory help bring this notion home. After the victory in Maine in 2012, Freedom to Marry conducted a poll that asked a series of questions about the impact of the campaign. It asked where people got their information about the campaign. In response, people overwhelmingly said television, advertising, and news coverage. But when they were asked what changed their mind, the number one answer was "a personal conversation with someone I trust." The second most common answer was "a conversation with a woman" in their family. For Wolfson, this last finding helped underscore that women are "effective change agents" in talking about marriage equality and are good at "moving people around them."

Wolfson recognizes that the time the movement had between losing in Maine in 2009 and the ballot initiative for 2012 gave Freedom to Marry and its local and national partner organizations the opportunity to take their refined message door to door and "make the conversation personal." This combination led to the victories in 2012 generally, not just in Maine. For example, Wolfson points to the fact that when President Obama came out in favor of marriage equality in 2012, he referred to the effect that his personal relationships had on his thinking on the subject. As Wolfson described it, Obama used the language of a "Christian and a parent," not the language of a "constitutional scholar."[11] The president talked about his family and teaching his children; he talked about his own state of mind and about conversations with his staff members and his family that ultimately led him to formally change his public stance on marriage equality. For Wolfson, this was a very powerful example of how people change their minds: "This is how we had built our majority, how millions of people have changed their mind." Wolfson said the movement accelerated as the president spoke out in favor of marriage equality.

At the national level, the campaign for marriage equality had several components: lobbying, litigation, and fundraising. For Wolfson, before the victory at the Supreme Court, what "propell[ed] things forward" were "the conversations" taking place at the local level. As a result, the campaign tried to find ways to encourage those conversations to occur. According to Wolfson, "that's what built the majority" the campaign had earned. It is also what resulted, as Wolfson had predicted, in "finishing the job." But before discussing the campaign's ultimate victory at the Supreme Court, I want to take a moment to briefly assess the Maine campaign to determine the extent to which it embodied the components of the social change matrix.

The marriage equality coalition in Maine connected to the social change matrix primarily by harnessing the power of its message. The campaign used unifying, personalized messages to connect a broad range of individuals to a broad campaign. By using the language of marriage equality, connecting to prospective voters, and developing trust on a personal level, the campaign seems to have perfected the art of messaging. A deeper look into the campaign shows that it seems to have harnessed the other components of the matrix as well: that of network and medium.

First, Freedom to Marry formed as a national organization that connected to local organizations that worked at the state level to promote marriage equality through a number of legal mechanisms. A typical approach for these campaigns was to have a state-based organization employ staff and recruit volunteers on the ground. National groups often served on a steering committee for a state-specific campaign. In Maine, Equality Maine existed before the campaign that culminated in the 2012 vote. However, the later campaign was organized through a new entity: Mainers United for Marriage (the name even suggests unity). Groups like

Freedom to Marry, the Human Rights Campaign, Equality Maine, and ACLU of Maine were represented on the steering committee of the campaign. The national groups offered technical assistance, particularly with research, messaging, and funding. They used tools such as conference calls and e-mail to communicate instantaneously throughout the country. The steering committee did not need to be on the ground in Maine to receive real-time updates on the progress of the campaign.

Second, technology helped facilitate the work of the campaign in many ways. Canvassers had iPhones and iPads as they went door-to-door. They could use these devices to show video clips to prospective voters. They could also access information about particular voters that might be in the campaign's database already, and they could input data about voters with whom they had met immediately. This eliminated the need for canvassers to input data when they returned to the office at the end of a long day, and it meant that the campaign could track voters in real time. It also gave the campaign insights into the types of conversations canvassers were having with particular voters and the messages that seemed to move them. Later, when it came time to motivate voters to go to the polls, the campaign could tailor and target particular messages to specific individuals based on what types of messages canvassers had identified as those that might influence such voters.

In addition, the campaign was able to circulate the stories prospective voters had shared through social media. Mello grudgingly appreciated the value of this medium. "As a field person," she admitted, "I am sometimes skeptical of social media." However, "the biggest benefit we got [from sharing stories from the field] was that we could build the energy and sense of momentum for our volunteers and donors." The campaign used social media platforms such as Facebook and Twitter to share stories of people who had changed their minds and decided to support marriage equality. Sharing this information not only built momentum for people to join the movement, it also helped shape public opinion more generally. It made the canvassing effort seem less intimidating for both volunteers and prospective voters. It also showed prospective voters that they were not being singled out for these conversations. Rather, such discussions were happening throughout the state and thousands were participating in them.

At the national level, Freedom to Marry harnessed technology to coordinate public information about the broader campaign, acting as the hub of advocacy and the social media campaigns that took place across the country. The organization's Digital Action Center website served to provide support to all the state campaigns. This work was "shaped by local stories, local people, local experience," says Wolfson. Echoing the efforts of Medgar Evers and his television broadcast to the residents of Mississippi during the civil rights movement, local advocates talked with local audiences to promote marriage equality. In communities across the country, this effort connected the public "to an array of stories, and voices, and messengers." Wolfson spoke of this as the "air cover" that supported the work being done on the ground, the thousands upon thousands of conversations that were happening across the country around these issues. The organization supplied the digital support to such conversations through Twitter, Facebook, and various websites, including one geared toward the Latinx community (familiaesfamilia.org) and a page on the Freedom to Marry website geared toward young conservatives. It also set up another site, whymarriagematters.org, which served as a repository of information for the values-based, personal framing and messaging that campaigns could utilize. In each instance, the information was made available for local campaigns and emphasized the values-driven message.

For Wolfson, this array of powerful tools was not around when he started the work in the 1980s. E-mail, the Internet, social media: this "constellation" of resources, as he describes it, had "become hugely significant' in the day-to-day of the campaign. "None of that existed" when he started his work, and some of it had "grown substantially" in previous years, he explained.

In these ways, the movement for marriage equality, as it was building steam and gathering support across the nation, tapped into the elements of the social change matrix to advance their goals. That effort led to victory at the highest level: landmark wins at the U.S. Supreme Court, the culmination of a decades-long struggle.

Roadmap to Victory

Freedom to Marry laid out what it called its "Roadmap to Victory." This roadmap, according to Freedom to Marry's website, called "for advancing work on three tracks—winning more states, growing the majority, and ending federal discrimination." The goal of this effort was to "return to the U.S. Supreme Court with a critical mass of states and undeniable momentum in public opinion, the conditions history tells us are required for the Supreme Court to be most likely to rule" in favor of marriage equality. For Wolfson, this strategy came "from the lessons of history." And the reason Freedom to Marry existed was to "propel this strategy and then go out of business."

Wolfson believed the final goal of the campaign was for the Supreme Court to bring the country to a "national resolution" on the marriage equality issue. He did not believe the campaign had to win in every state, just enough states that it could build enough support to create a climate that convinced "the judges and ultimately the justices to do the right thing." Although he did not know the timeline when I spoke with him in February of 2014—whether it was "months or years"—Wolfson believed at the time that the U.S. public was ready for the Supreme Court to rule in favor of marriage equality. What the campaign needed to do was "convey" to the courts this fact: that "America is ready."

Recent longitudinal research on the changing public attitudes around marriage equality from the late 1980s to the mid-2010s shows that approval of same-sex marriage changed dramatically during this period, likely the result of a broader cultural shift happening in which it first became more commonplace for people to self-identify as being lesbian or gay, a product of shifting attitudes more generally, but also courage on the part of those who came out. This research shows that people who stated that they knew members of the LGBTQ community were more likely to support marriage equality.[12] So while the prospect of legal change was unfolding, public opinion was also shifting, and such broad public opinion can often have an effect on the ways in which the Supreme Court can reach decisions on issues of broad public import.[13] In Wolfson's view, the campaign would win once it was able to place "the right case in front of the right justices at the right time." Although in many ways this could not be controlled, the campaign needed to focus on those aspects of the litigation that were "in our control," says Wolfson. For him, the "end game" was victory before the U.S. Supreme Court.

When asked whether advocates for marriage equality might face backlash should the Supreme Court rule in favor of marriage equality, Wolfson responded that he did not believe victories in the courts for marriage equality would result in any sort of retrenchment or reversal of rights, like that which has been seen in other contexts where courts have entered the fray. For him, the stakes regarding marriage equality are different from those in other contexts where the courts have established rights where people's emotions may run strong. "Nothing actually bad happens, no matter what you think about" marriage equality. "You just may not like it, but nothing bad happens," when lesbians and gays attain equal marriage rights, he said.

For Wolfson, the best analogy to what might happen in the wake of a national, pro-marriage equality ruling is what happened after the Supreme Court found that state laws banning interracial marriage were unconstitutional: *Loving v. Virginia* (which Wolfson describes as "the best-named case ever"). "When we won the end of race discrimination in marriage," Wolfson says, "there were no riots, there were no southern manifestos of resistance, there were no bombings." The stakes here are no different. In Wolfson's opinion, no one actually loses anything. Opponents of marriage equality are "increasingly isolated and dwindling," Wolfson says.

Justice Ruth Bader Ginsburg made waves when she expressed her belief that the Supreme Court's action on abortion rights in the early 1970s helped provoke a backlash against such rights, and that perhaps the court had acted hastily, before the country was ready for such a broad and sweeping recognition of a woman's right to reproductive freedom. "It's not that the judgment was wrong, but it moved too far too fast," she is reported as having said at a conference at Columbia Law School.[14] Wolfson disagreed with Justice Ginsburg's take on the aftermath of *Roe v. Wade* and he did not think there would be any national backlash from a pro-marriage equality decision out of the Supreme Court: "There won't be the kind of post-*Roe* struggle that Justice Ginsberg seems to be talking about, rightly or wrongly."[15]

For Wolfson, the environment has changed from the early days of the marriage equality movement. Back then, in Wolfson's words, "when we won anything, there was immediately a dip in support." In the last several years, however, there has been bipartisan support for marriage equality, "giant wins" in 2012, and President Obama's expression of his support. In the wake of these events, there have been no bad effects or no weakening in support for marriage equality. Quite the contrary, Wolfson said, "people are embracing it." Unlike in 2004, when Republicans tried to get marriage equality bans on state ballots as a way to drive voter turnout, in more recent years, Democrats tried to put the issue on state ballots, to do the same, just to get different voters there.

Wolfson and the Freedom to Marry team saw the movement as consisting of many voices and many groups, all working together. Indeed, the legal team, or teams, that would litigate a wide range of cases, all pressing toward the ultimate goal of victory before the Supreme Court, included lawyers from the American Civil Liberties Union LGBT & HIV Project, Lambda Legal, a range of other nonprofit groups, countless volunteer lawyers, and movement lawyers from private law firms, like Roberta Kaplan.[16] For Wolfson, as long as everyone was furthering the broad strategy, Freedom to Marry welcomed and supported everyone's varied contributions. It was a "movement," in Wolfson's words, "and not an army." The goal was to make the whole "greater than the sum of the parts."

In late June 2015, Freedom to Marry and millions of supporters got their wish. In a 5–4 opinion, the Supreme Court declared that laws prohibiting same-sex couples from marrying were unconstitutional. In other words, all states had to recognize marriage equality: lesbian and gay couples throughout the United States would have the same right to marry as heterosexual couples. In the decision, Justice Anthony Kennedy, writing for the majority on the court, borrowed from the Freedom to Marry playbook. The majority opinion referenced the personal stories of several plaintiffs in the case, like that of a sergeant in the U.S. Army Reserve, Ijpe DeKoe, who married his partner in New York, where same-sex marriage was legal, prior to DeKoe's military deployment in Afghanistan. On his return, he and his spouse moved to Tennessee, where their marriage was not recognized, "disappearing as they travel across state lines," as the court wrote.

The opinion also confirmed the message that the marriage equality movement had begun advancing in earnest since the 2008 loss in Proposition 8, and which Wolfson and Andrew Sullivan had been promoting for years. The campaign was not about attacking so-called traditional marriage; it was about affirming it. Gay and lesbian couples wanted to join the institution, not tear it down. As the court wrote:

> No union is more profound than marriage, for it embodies the highest ideals of love, fidelity, devotion, sacrifice, and family. In forming a marital union, two people become something greater than once they were. As some of the petitioners in these cases demonstrate, marriage embodies a love that may endure even past death. It would misunderstand these men and women to say they disrespect the idea of marriage. Their plea is that they do respect it, respect it so deeply that they seek to find its fulfillment for themselves. Their hope is not to be condemned to live in loneliness, excluded from one of civilization's oldest institutions. They ask for equal dignity in the eyes of the law. The Constitution grants them that right.[17]

What also seemed to resonate with the justices in the majority were the state-by-state efforts to support marriage equality: at the polls (as in Maine), in state legislatures, and in state courts (like in Hawaii and Massachusetts). The majority opinion would explicitly reference, and catalog in an appendix to the opinion, evidence of what it describes as "years of litigation, legislation, referenda, and the discussions that attended these public acts."[18] That the momentum seemed to be building in the states to recognize marriage equality appeared to have had some impact on the court's decision. So efforts like those in Maine and other states, just as Wolfson and many others had hoped, seem to have had some impact on the court's ultimate decision.

While it is often hard to tease out what moved each justice to reach a particular decision, one thing is certain. The message that Wolfson promoted while still in law school seems to be one that courts, state legislatures, and voters ultimately moved toward. In that paper, he said:

> People are born different, into different circumstances, but are inherently equal in moral terms and in the eyes of the law, as our Constitution confirms. According this equality is perhaps most vital when it comes to love, the great leveler, which comes to each of us not wholly by choice or design. The choice we do and should have is what to make of what we are. For gay women and men, who also love, same-sex marriage is a human aspiration, and a human right. The Constitution

> and real morality demand its recognition. By freeing gay individuals as our constitutional morality requires, we will more fully free our ideas of love, and thus more fully free ourselves.[19]

The Supreme Court ultimately embraced that law student's vision, establishing it as the law of the land. The court found that the marriage "dynamic allows two people to find a life that could not be found alone, for a marriage becomes greater than just the two persons. Rising from the most basic human needs, marriage is essential to our most profound hopes and aspirations."

That paper Wolfson wrote, perhaps one of the most prescient and artful ever written by a law student: he got only a B.

8

A LIVING WAGE IN LONG BEACH

Sandwiched between the heart of Los Angeles and Orange County, California, Long Beach is a port city of nearly five hundred thousand residents. Traditionally a home to manufacturing and a thriving commercial port, the city has seen a reduction over the past half century in the number of well-paying manufacturing jobs at companies like Boeing. In many ways, Long Beach is a microcosm of the growth of economic inequality in the United States over the last forty years. Once home to thriving aerospace industry manufacturing, which paid strong, middle-class, often union wages, the city's fortunes changed in the late 1970s, as defense contract cutbacks, and the shifting of jobs overseas and to states where employers could pay a lower wage, weakened the city's middle class. In 1978, in the throes of California's tax rebellion that made local property taxes much more difficult to raise, Long Beach was described in a federal government report as one of the most financially distressed cities in the U.S.[1] Yet by 2007, the Urban Land Institute named the city as having one of the country's top ten revived downtowns.[2] What brought about this shift in fortunes? How did the city turn its prospects around over the span of just a few decades?

Although Long Beach's port has remained a viable hub of commerce, one of the biggest shifts in the economy of the city has been a movement from manufacturing jobs to service jobs, mostly in the hospitality industry. Although downtown Long Beach has been revived, the Brookings Institution identified Long Beach as having the sixth-highest concentration of low-income people in the country.[3] Indeed, unfortunately for Long Beach's low-wage workers, the shift to a service economy has meant jobs that pay a lot less than those in manufacturing.

What helped bring about Long Beach's economic turnaround, shifting from well-paying manufacturing to lower-paying hospitality jobs? One cause may be the city government's policies. Since the early 1980s, Long Beach's municipal government has offered over $100 million in tax breaks to hotels to induce them to locate or stay in the city and has spent hundreds of millions more in infrastructure improvements and amenities, like remodeling the city's convention center, guaranteeing the bonds for the construction of an aquarium, supporting the construction of a retail and entertainment complex, revitalizing the downtown through improvements to sidewalks and other amenities, and providing water taxi services.[4] The city targeted these efforts toward boosting the tourism industry, benefiting existing hotels and businesses the city was able to lure to the area. All in all, the city of Long Beach has spent roughly $750 million to bolster the hotel industry through direct subsidies and by making the city an attractive destination for tourists and business travelers. However, many of the workers in these hotels earn no more than the state's minimum wage. As a result of these trends, Long Beach has become, in the words of one local advocacy group, "two cities:" one wealthy and one poor.[5]

That group, the Los Angeles Alliance for a New Economy (LAANE), studied the tourism industry in Long Beach and highlighted both the role of tax breaks for this industry in the community as well as the low wages paid to workers in it. LAANE teamed up with a local chapter of UNITE HERE Local 11, the union that represents workers in the hotel, food service, and gaming industries, and began to advocate for higher wages for employees in hotels in Long Beach. LAANE issued a report on these issues that helped expose some of the economic issues facing low-wage workers in the hotel industry.[6] Some of the findings of that report were that 40 percent of Long Beach's workers earned less than twice the federal poverty line, meaning they likely did not earn enough to survive without government subsidies to make ends meet. Indeed, a survey of workers in one hotel revealed that 41 percent of the workers there utilized some form of public assistance, like food stamps or the state's version of Medicaid. LAANE's analysis of workers in this hotel revealed that their incomes qualified them for an average of $2,735 in public assistance per year. LAANE found further that more than 18 percent of Long Beach's residents were below the federal poverty line, a population that included 25 percent of the city's children. The median family income in Long Beach was 10 percent lower than the median income in Los Angeles County and 19 percent lower than California as a whole. Hotel workers earned an average of roughly $19,000, less than one-third of what workers in manufacturing earned and less than half the median income in the city. The report also attempted to put a human face on the issue of low wages in Long Beach by including pictures of workers and stories of individual workers. Most important, perhaps, it explained the economic hardship workers faced even when they were working full-time.

To address these issues, LAANE joined forces with the Local 11 to try to strengthen the wages of Long Beach's hotel workers using a strategy that progressive cities across the country are beginning to adopt. This strategy—raising the minimum wage of certain sectors, or in states, cities, and localities as whole—is proving an effective way to raise the incomes of low-wage workers. Can groups advocating to raise the minimum wage incorporate elements of the social change matrix when doing so? This coalition's work just might reveal a path forward for such efforts.

While the local city council in Long Beach is largely made up of Democratic elected officials who might be sympathetic to the cause of low-wage workers in the hotel industry, the coalition sensed that the council would not pass an ordinance increasing the minimum wage for hotel workers because of the political strength of the hotel industry in Long Beach. Council members told coalition representatives that while they personally supported a higher wage in the hotel sector, their constituents were more conservative and thus they would not vote in favor of such legislation.

In light of their fear that they could not get an ordinance passed through the local city council, the coalition began to explore the possibility of putting a referendum on the ballot in Long Beach in November of 2012. The referendum they promoted sought to raise the wage of workers in hotels with more than one hundred rooms from the state's minimum wage of eight dollars an hour to thirteen dollars an hour. It also provided five paid sick days a year and ensured that when a hotel hosted a banquet it could not pocket the fees it was passing on to customers that many believed would go to the workers as gratuities. Hotels had a practice of adding "service" charges to the cost of banquets (often as much as 20 percent), but workers did not receive that money. The hotels typically kept it.

California law gives local governments some control over pay and conditions of employment for workers and gives voters the power to pass referenda on such issues. The UNITE HERE-LAANE partnership learned from its own polling that the proposal to raise the minimum wage of hotel workers had broad support from across the political spectrum, including Republicans and Democrats, men, and especially women. Armed with this knowledge, the partnership set out to build a broad-based coalition that would support a ballot referendum to raise the wages of workers in larger hotels in the city. They sought the likely allies (tenant associations, liberal congregations, LGBTQ activist groups), but they also sought out unlikely allies in ingenious ways that brought the social change matrix to life. The coalition was inspired by a credit card company's advertising campaign,

"Small Business Saturday," which encouraged consumers to "buy local." The leaders of the partnership had the idea of emphasizing that an increase in the minimum wage for hotel workers, many of whom actually live in Long Beach, would give them more disposable income that they would spend at local businesses. The campaign began an effort to enlist the support of area small business owners who stood to benefit from an increase in disposable income for local workers. As Leigh Shelton, who was the Local 11 communications director during the campaign, described it, "We didn't need to present it as some revolutionary idea. It was already something people felt positive about and felt good about, so we wanted our materials and our message to remind them about how good they feel about raising the minimum wage" for hospitality workers.[7]

The referendum came up for popular vote in the fall of 2012, a presidential election year. For Shelton, in that year, "reasonable was the name of the game." People were just climbing out of the recession and wanted to support reasonable policies and elect reasonable leaders. The recession was over in Long Beach's hospitality industry, but the increased earnings were not trickling down to the local level. Although the economy was improving, the coalition's assessment was that Long Beach residents did not have enough money to spend at the local level. For Shelton, spending local was a "common sense message" that made people feel good: "People like to shop local; they like to be able to go to a store down the street and spend their money there, especially when they know it is run by a member of their community." The campaign to raise the minimum wage was not just "a lefty social justice issue that only liberals care[d] about." Jeanine Pearce, a community organizer at LAANE who was elected to the Long Beach City Council in 2016, believes that this message was critical. She described it as "talking to people the way they wanted to be talked to."[8] The coalition sought out grassroots leaders to carry this message and "change the conversation," in Pearce's words. She said that the workers were at the center of the campaign; they were the "champions." But the coalition also enlisted students, volunteers, paid staff, local residents, and local small business owners.

The coalition began the campaign by engaging in leadership development and training. They initiated a civic engagement program in which volunteers knocked on doors throughout the city, talking to people about supporting the effort. They also tapped into different demographic groups to find the right people to give the right message to the right audiences. For example, the coalition recruited a group of Filipino students as volunteers. After they were trained, these students spoke to members of the Filipino community in Long Beach about the ballot measure. Similarly, coalition members tapped small business owners and supportive residents who lived in more affluent parts of the city to reach out to other networks they might be part of, like a homeowners' or neighborhood association. The messenger spreading the word about Measure N, as the referendum came to be known, thus came from the same community as the target audience.

The organizers feel that much of the success of the campaign came from this practice of matching individuals from a neighborhood, group, or community to deliver messages that would resonate with their friends and neighbors. In low-wage worker communities, a message about raising wages for hotel workers would have impact because most residents of those neighborhoods had a relative, friend, or neighbor who worked in the hotel industry who would benefit from higher wages. In more affluent communities, the idea that local small businesses would benefit if local residents had more disposable income proved effective, because the community as a whole would improve. In communities that might be more conservative politically, the campaign's representatives emphasized that research showed that for every dollar a worker makes in the city, three dollars go back to small businesses.

An important task of the coalition was finding out what community connections their supporters had. When the campaign had identified a resident of a particular community who had relationships with others in that community, that individual would go speak to his or her own neighbors about the effort. Through such channels, a neighborhood association in a community whose residents might not be expected to want to help workers would have one of its members speak in favor of the ordinance. Pearce says that people were "a lot less leery" of a neighbor than they might have been of an outsider. They also sought unlikely allies, such as environmental groups, to broaden their base of support. As Pearce puts it, someone might say, "I can't be against this" if a group that person was affiliated with had publicly expressed support for the effort. The campaign thus tapped into preexisting social capital ties to expand its reach and support.

Pearce reports that identifying the right message and having the right messenger deliver it, often "through the relationships that people had," was key to the success of the campaign. The coalition made a variety of appeals that were tailored to specific communities and were delivered by carefully selected messengers. Its positive message seemed to resonate across all communities: "Everybody likes to feel good about a positive campaign," Pearce says. This made it easier for campaign members to garner support from residents and to recruit volunteers and keep them motivated and energized about the effort.

Pearce reports that the "design and feel of the campaign," was positive and nonthreatening. In the end, the message that "we're all going to do better" if workers have more money to spend resonated across all demographic groups in the city. This message stressed interest convergence: more money for low-wage workers meant more money would go back into the community and that would benefit everyone. The campaign had made Measure N a "moral issue," but that morality was not necessarily about workers. A strong piece of the argument was that it was about small business owners. Pearce says that "the message was key and we continued to fine tune it every single day."

As with the marriage equality effort in Maine, the campaign's outreach involved engaging in substantive conversations with small business owners to enlist their support. While some business owners worried that the broader goal of the campaign was to raise the minimum wage across the city, the organizers made it clear that the effort was only directed at the larger hotels, the ones "who could handle" an increase in wages. Support from the small business community turned out to be strong, and the campaign enlisted two hundred small business owners to endorse Measure N. "Yes on Measure N" posters began appearing in stores that supported the effort.

The campaign also began an ad campaign promoting positive messages and featured photographs of workers spending money in local area stores. As one flier put it: "We work here. We live here. We spend here." The photographs typically featured a smiling worker putting money on the counter of a local store. "We had pictures of people who were happy, with cash in their hands," recalls Loraina Lopez Masoumi, an organizing director at UNITE HERE. "It was an opportunity to have that worker, a minority worker, with a very happy face with cash in their hand, giving it to the seller, or on the counter. This was very important in the messaging: cash and happy faces."[9]

At first the campaign spent a lot of time and money sending canvassers door-to-door to collect enough signatures to get the referendum on the ballot. During this push, campaign workers also tried to have conversations with residents about the importance of the ballot initiative. When it was time for the campaign to get supporters out to vote on the measure, residents remembered these previous conversations and they often reported that they felt a connection to the campaign. That was the goal, Shelton recalls, to "increase opportunities for the voter to feel connected to the worker." When it came time to vote, many residents were happy to support the referendum.

Shelton says that the organizers wanted campaign materials "to have that [personal feeling] too. . . . We wanted our materials to have that same spirit." As the LAANE report on Long Beach had done, the ballot campaign used the images and stories of real individuals in its advertising efforts to personalize the campaign so that voters could link the face of a local worker or small business owner who might be a neighbor or a friend with the campaign. The images were also diverse and featured Asians, Latinos, African Americans, and whites.

Shelton recalled that the campaign "always looked for ways to create a human connection, even if it [was] through a mail piece." The organizers introduced many individuals and stories as a way to find what might stick with a particular audience. "You never know who people are going to feel connected with, or what part of someone's story" will resonate, "so we tried to have a variety of spokespeople, from the hotel workers themselves to the small business owners." Pearce said that using actual Long Beach residents in the campaign gave it a "very homegrown," feeling. Many people knew a worker, a neighbor, or a small business owner who was connected to the campaign.

The UNITE HERE-LAANE partnership did not do much in terms of social media to promote its message, although it had a Facebook page and used Twitter and other channels. Instead, it used the mails and, most important, the face-to-face, door-to-door canvassing to get its message out. On the day the nation reelected President Obama, Long Beach voters passed the measure by a 2–1 margin. The large majority has led some leaders to wonder whether they could have sought an increase of as much as eighteen dollars an hour.

The measure has not been without its costs, however. Some hotel owners have retaliated by demanding more productivity from workers and imposing new and different work requirements, insisting that they need to make up for the higher wages they are paying. Two hotels even reduced the number of rooms they offer to guests to fewer than one hundred so that the living wage ordinance does not apply to them. Because the hotels that ratchet up the job requirements on employees are often nonunion sites, union leaders emphasize that the best way to prevent such retaliation is to increase the unionized work force. Because of this, they are always looking for opportunities to expand the number of unionized hotels in Long Beach, which is no small feat, but one the union pursues wherever there appears to be an opening.

In the end, though, as Shelton said, "the sky didn't fall." Instead, there was a new spirit in the community. The Long Beach campaign "shifted the dynamic so much," Pearce said. Winning so convincingly at the ballot box meant that everyone realized that a supermajority of Long Beach residents supported these issues, and city council members no longer had to worry that their constituents did not support these sorts of issues. Pearce said that "Long Beach is forever changed because we . . . have empowered people to think outside the box, and to push policy and to push politics and to push the discussion." In Long Beach, based on the network they created, "now we can say we've got twenty organizations that are doing real organizing." The most important thing to Pearce was that groups in the city feel empowered, hopeful, and confident that they can launch more successful campaigns around similar issues. One of the key lessons Pearce learned from the Long Beach effort was that one can "never write anybody off." In addition to talking to usual allies, the campaign "spent a lot of time speaking to Republicans." This was part of a strategy focused on "getting to scale and getting to the tipping point." To do this, the campaign focused on face-to-face interactions and making other personal connections.

While Pearce would like to see an increase in the federal minimum wage, in her opinion, a state-by-state approach stands a better chance of success. At the same time, she feels that city-level organizing can benefit from personalized interactions in ways larger campaigns cannot. Through these more localized, targeted efforts, campaign workers can really "get out and talk to voters." In Pearce's analysis, "if you don't have that level of 'I'm your neighbor and here's my story,' and connect with people, then the campaign is always going to be about someone else, and, unfortunately, the majority don't do things for somebody else that they don't know. . . . That's why the personal part is important."

Shelton explained that the campaign resonated with people for many reasons: because it had a positive message, because it was targeted and people could see how Measure N would benefit them, and because there seemed to be a desire among people across the political spectrum to do something about income inequality. Shelton noted that "income inequality is a dire problem that needs fixing" and felt that people were looking for initiatives and measures to address it. In her opinion, this campaign gave them just such an opportunity.

The success of Measure N reveals the social change matrix at work. The UNITE HERE-LAANE partnership advanced a unifying, positive message that tapped into the convergence of interests among low-wage workers and small business owners—two groups that are not typically allies on efforts such as this. It also personalized the message by identifying individual workers who lived and worked in Long Beach and stood to gain from a wage increase. Emphasizing the fact that workers lived in the city reduced the social distance between them and more well-to-do voters. The campaign tapped into notions of shared humanity and shared destiny, ideas that have been the hallmarks of many successful social change movements in the past.

After their success in Long Beach, LAANE and UNITE HERE successfully promoted a higher minimum wage for certain workers in the hotel industry in Los Angeles. In 2014, UNITE HERE won a $15/hour minimum wage for hotel workers in larger hotels in that city. The new campaign was designed to phase in an increase in the minimum wage for hotel workers based on the size of the hotel: hotels with greater than three hundred rooms had to pay $15.37 an hour to their workers by July 1, 2015, and hotels with greater than 150 rooms had to pay that amount by July 1, 2016. Subsequent advocacy raised the minimum wage in Los Angeles across the board, not just for hotel workers, but not as high as the hotel workers were able to secure.

LAANE and UNITE HERE took several pages out of the Long Beach playbook when the joint group promoted a living wage ordinance for hotel workers in Los Angeles. They published a report on the topic, as they had done in Long Beach in advance of the vote on Measure N. In that report, they emphasized that increased pay for low-wage workers would translate into more money going into the local economy. They estimated that the wage increase would put roughly $73 million more in the pockets of hotel workers, of which $41 million would be spent in the local economy. They also estimated that this would generate an additional $30 million in increased economic activity. The report used the personal stories and pictures of a worker and a small business owner, both of whom emphasized that a higher minimum wage for LA's hotel workers would benefit the local economy. Learning from their experiences in the Long Beach campaign, advocates used elements of the social change matrix such as emphasizing interest convergence and connecting with local residents on a personal level.[10]

One question that efforts such as the Long Beach campaign raise is whether increasing the minimum wage is a viable tool for combating economic inequality. Critiques of such efforts often raise concerns that raising wages will have the unintended effect of reducing the number of jobs available, that employers will cut back or go out of business altogether because of increased labor costs. If such concerns are valid, then raising the minimum wage is ineffective as a tool for reducing economic inequality. Fortunately, the question of how increases in the minimum wage affect the number of jobs has been the topic of dozens of studies over the last half century.

Most of these studies show a slight negative effect on employment when the minimum wage in an area increases. Most of this impact falls on teen workers, but even there the impact is practically negligible. Congress commissioned one of the first of these studies when it created the Minimum Wage Study Commission in the 1970s. That commission looked at the estimated impact of pegging the minimum wage to inflation and raising the minimum wage at a lower amount for teens.[11] Economists reviewing the findings of this report, while synthesizing the results of other studies as well, found minimal impact on employment for age groups other than teens, and they estimated that teen employment might be reduced, at most, by just 1.5 percent.[12]

States are free to set a minimum wage that is higher than the federal minimum, and in many states, towns and cities can set a minimum wage that is higher than the minimum in their state. As a result, differences between minimum wage rates in states and cities can create an environment in which to study the impacts of raising the minimum wage. It is relatively easy to compare the impacts between similar geographic regions that have different minimum wage laws. Such studies provide a clear picture of how these changes affect local economies.

One of the most influential of these types of studies took place in the early 1990s, when David Card and Alan Krueger studied the impact on employment of an increase in the minimum wage in New Jersey. Pennsylvania had not increased its minimum wage at that time, and they were able to test the impact on employer behavior of a minimum wage increase in fast food restaurants in New Jersey and compare them to employer practices in neighboring Pennsylvania. Their study revealed that the increase in the minimum wage did not bring a larger decrease in employment in fast food restaurants in New Jersey than in Pennsylvania.[13]

Since the Card and Krueger study, studies of the employment effects of increases in the minimum wage have numbered in the hundreds. A metastudy conducted by John Schmitt of the Center for Economic and Policy Research that compiled the findings of decades of research reached the conclusion that the negative effect on employment of increasing the minimum wage is negligible. However, Schmitt found that while employers may not reduce the number of employees they hire because of an increase in the minimum wage, they may reduce nonwage benefits. In addition, an increase in the minimum wage may reduce company profits. But his study also finds that a minimum wage increase can lead to other economic benefits, such as increased productivity: workers likely expend more effort when they are paid more. In addition, employers may experience less turnover because workers who are earning more may stay in their jobs longer. Lower employee turnover means lower costs associated with searching for and training new employees. These costs can be significant.[14]

Although these studies seem to indicate there is little or no reduction in employment if the minimum wage is increased, can an increase in wages impact economic inequality? We know that through the late 1970s, a relatively higher federal minimum wage was in place, and when that started to decline in real value, inequality grew.[15] What is more, increases in the minimum wage over time have not kept pace with inflation. In 1968, using 2013 dollars, the minimum wage was, in effect, $10.69 an hour. By 1979, it had dropped slightly to $9.67. By 1990, it had fallen to $6.84.[16] One study of the decline in the real value of the minimum wage from 1979 to 1988 found that this accounted for roughly 25 percent of the income inequality among men and 30 percent of income inequality among women.[17]

A study by Arin Dube, a prolific minimum wage researcher at the University of Massachusetts Amherst, shows that a 10 percent increase in the federal minimum wage would reduce the poverty rate by roughly 2.4 percent and that low-wage workers would benefit the most.[18] In 2012, economists at the Economic Policy Institute estimated that a phased-in increase in the federal minimum wage to $9.80 an hour would help 28 million workers, who would receive nearly $40 billion in additional income. Most of these workers would be in households earning less than $60,000 per year.[19] As in Long Beach and Los Angeles, community organizing has won hard-fought victories to raise the minimum wage for workers in cities, such as Seattle, and states like New York.

As this book goes to print, groups are leading a "Fight for $15"—a fifteen dollar minimum wage. Studies tend to show that such an effort would boost wages and have few negative effects.[20] Since the growth in income inequality over the last forty years seems to correspond to the gradual reduction over time in the real value of the federal minimum wage, increasing the minimum wage would seem to be one very direct way to increase earnings, particularly at the bottom end of the economic spectrum. Other suggestions for how to combat economic inequality include increasing tax rates on the wealthy and strengthening educational opportunities. In his 2014 book *Capital in the 21st Century*, the French economist Thomas Piketty has suggested a "global wealth tax" that would help balance out some of the global inequities in wealth and root out tax havens where the wealthy are able to shield a percentage of their assets.[21] Others advocate for a universal guaranteed minimum income.[22] Without debating the relative merits of these different strategies, if the social change matrix is to be deployed in the service of combating income inequality, several approaches should be considered that have the potential to garner broad support. Minimum wage campaigns designed in the way that worked well in Long Beach, which harnessed the elements of the social change matrix, might lead the way toward addressing economic inequality, in towns, cities, and even the nation, moving forward. In the next chapter, I explore and assess what the social change matrix might offer advocates seeking change in this and other areas in an attempt to understand what could be the future of change.

9

PUTTING THE MATRIX TO WORK

When it finally passed, the G.I. Bill granted up to a year of education and training to veterans who had served in the military for at least ninety days, with additional benefits based on service time, up to a maximum of forty-eight months. Fifty-one percent of World War II veterans—7.8 million individuals—used the education and training benefits of the G.I. Bill, and by 1947, roughly half of all students enrolled in U.S. colleges were recipients of G.I. Bill education benefits.[1] By 1955, the cost of the G.I. Bill's education and training program totaled $14.5 billion—or the equivalent of more than $128 billion today.[2] As discussed in previous chapters, the American Legion seems to have tapped into three components of successful social change when it advocated for passage of the G.I. Bill. Not only did its leaders utilize the best available communications technology with great effect—from the telegraph to the movie house—they also worked through a vast translocal network of chapters spread throughout the U.S. They then promoted a message of inclusion, one that hinged on the personal connections people felt toward the returning veterans, a product of the media advocacy that was designed to evoke sympathy for them, some of whom at least were identified through the personal stories the Legion would promote in its advocacy.

From the postal system in the early republic, to the television in the mid-twentieth century, social movements used technology to build trust, foster collaboration, and spur social change. Whether it was the revolutionaries who embraced the printing press, abolitionists who harnessed steam, or civil rights advocates who used the television to their advantage, the emergence of new technologies seems to create opportunities for new social movements to gather supporters, spread their message, and advance social change. But technology is not destiny; it alone does not determine how humans will use it, and how social movements will put it at their disposal. Indeed, at the dawn of the Internet age, the sociologist Manuel Castells argued that "many factors, including individual inventiveness and entrepreneurialism, intervene in the process of scientific discovery, technological innovation, and social applications, so that the final outcome depends on a complex pattern of interaction."[3] I have argued that recent advances in communications technology mean we are in a social innovation moment at present: a time when such new technologies may create an opening for new social movements to emerge. What the history of the last social innovation moment reveals, when computerized mailing lists helped foster the emergence of new types of organizations, is that the ultimate impact of technology on social movements is not foreordained: it will not lead inexorably to progressive social change. If history is any guide, social movements will need to harness contemporary technologies, but also do other things well, to bring about such change. In this book I have tried to show how social movements can thrive in those moments when new forms of communications emerge. Often in these moments, social movements utilized the most modern technology available to organize and communicate; created different types of organizations; and attempted to address social inequality in different forms, from the political inequality in the colonies to the racial inequality of mid-twentieth-century America. What follows is an attempt to synthesize the more distant history of social movements in the United States with the case studies of contemporary efforts that have used the new tools of communication to sketch out a playbook for social change advocacy in the current social innovation moment.

Using Medium to Connect, Coordinate, and Lower Social Distance

A central lesson of some of the more successful movements in social innovation moments is that they have all made effective use of the most modern forms of communications technologies to form, organize, and communicate their messages and coordinate their constituents' activities in an effort to advance social change. That is evident from the colonial era to today. Whether it was the printing press, the television, or the telegraph, social movements have often arisen in tandem with advances in communications technology, and often, in these movements, social movements formed translocal networks. In contrast, however, in the early 1970s, a new medium helped dictate the shape of social movements for two generations. The ability to create computerized databases and to form national organizations without the benefit of true, translocal networks ushered in a new era of advocacy, one marked by national organizations that were, for the most part, unconnected to local groups. Contrast that with more contemporary efforts—the advocacy of the West Virginia teachers, the campaign for the expansion of VAWA, the marriage equality movement, and Long Beach's living wage effort. These contemporary efforts harnessed modern communications technologies in ways that are markedly different from the movements that took root in the 1970s, returning to translocal, networked, face-to-face organizing and movement building.

For example, these contemporary groups used the latest forms of media in creative ways, ways that build trust, foster personal connections, harness grassroots support, and promote a positive message. In short, they used the power of the contemporary media at their disposal as a way to connect the different elements of the social change matrix as I have described them here. Where national advocacy groups in the recent past might have used the best technology available to them to build a membership base of individuals who wrote checks to those organizations to sustain the efforts of their leaders, some contemporary advocacy groups are changing the dynamic, connecting in ways that foster interpersonal connections while building broad-based support. They do this by emphasizing interest convergences, networks, social capital, and the creation of personal bonds. These approaches are very different from the ways that national advocacy organizations functioned from the late 1960s and early 1970s through to the mid-2000s. Utilizing a new communications technology—the computerized mailing list—they created a new kind of network, one that was centralized, with little connection between different members of the network at the grassroots level. This made the development of personal bonds between members difficult. Groups that consist mostly of members in name only, tied together by little more than a mailing list, find it difficult to form strong, face-to-face bonds between their members at the grassroots level. Sometimes they do not even know how to do so. As one example of this phenomenon, in the 1970s, when an in-house study of the field operations of the group Common Cause unearthed a desire for individual members to meet one another, such members were instructed to gather local members together in the following way: host a viewing party to watch the organization's leader speak on national television.[4] The group did not seem to have any tools for engaging their local members in effective ways that would help them communicate directly with one another and build those bonds of personal trust essential to identity formation and activism.

Just as previous advances in communications technologies and processes may have re-created themselves in the structure of those systems—think of the groups that arose in the early nineteenth century that were a reflection of the postal system—today's technologies are creating new kinds of organizations as well. Where hierarchical groups like some of the more prominent environmental organizations have structures that reflect that they are built on computerized mailing technologies, and now e-mail lists, newer organizations have the capacity to be much "flatter": they can be less hierarchical and enable all members of a group to communicate and coordinate with each other simply and systematically through electronic communications and social media. Less hierarchy means there is less social distance between leaders and members, and, more importantly, between members and other members. This diminished social distance can help lead to even greater trust within organizations. Technology is thus enhancing groups' capacities for building trust not just because it is creating less hierarchical organizations, however. It is also fostering the types of communications that help engender trust.

One of the groups that is at the center of this new approach is the Opportunity Agenda, a self-described "social justice communication lab" that works with both local organizing groups and national entities to craft progressive messages to have broad support. I spoke to Alan Jenkins, who led this organization for over a decade before leaving to join the faculty at Harvard. For Jenkins, technology is making these flatter organizations possible. He says that twenty-first-century movements are much different from those of the twentieth century, mostly as a result of technology. The latter were centralized and organized around charismatic leaders. Over the last decade there have been groups like the Occupy Movement, the Tea Party, the Wisconsin worker movement, the marriage equality campaign, and groups of immigrant youth. These are all diffuse and democratic. They are not centered on a charismatic leader. They are, according to Jenkins, "highly networked and highly keyed in to modern technology." In fact, for Jenkins, there is a "symbiotic relationship" between movements that are decentralized and the technology that facilitates that type of structure. Indeed, "there would have been no Occupy movement without social media," Jenkins asserts. "The medium shaped the movement itself."[5]

In the twentieth century, Jenkins asserts that groups were striving for what he calls their "Walter Cronkite moment": when the whole nation gets the message. That does not exist anymore, but communications strategies of late twentieth-century organizations were built around that model, when professional journalists with dedicated platforms and beats dominated the media. Now journalism is much more democratic and accessible. For Jenkins, this has profound effects for organizations and organizing. He asserts that it is the "first moment in human history where most of us at least in the U.S. can reach millions of people and they can reach back and we don't even know what that is going to mean for social movements as a practical matter." At a minimum, organizing can have a much quicker impact on others. The Arab Spring almost instantaneously spurred the Occupy Movement. Compare this, Jenkins states, with the nearly two decades it took for Gandhi's teachings to inform Martin Luther King, Jr., and the civil rights movement.

The speed and ease of this type of communication causes challenges, however. "There are lots of people who would not under any circumstances go out and protest at Zuccotti Park," Jenkins asserts (the park where Occupy Wall Street gathered). Those same people, however, might "spend hours in their home spreading information and connecting people and advocating for change." For Jenkins, this represents an expansion of a movement's base, "because those are people who, prior to the current era, there was no way that they were going to connect with social change." These individuals, says Jenkins, "might have been capable of it, but they weren't going to do it." At the same time, there were likely those who might have come out to protest, "but felt that they didn't have to because they could just 'retweet' occasionally." Jenkins does not know what the ratio is of the first group to the second group. Nevertheless, new media can tip the scales in favor of engagement, mobilization, and social change. The Opportunity Agenda now focuses on supporting grassroots and other efforts by helping them craft winning messages that are broadcast on different media, old and new. For Jenkins, just as technology has helped shape twenty-first-century advocacy organizations, "the medium shapes the message" as well. It is too early to predict the full scope and impact of these technological changes on organizing and organizations. Indeed, according to Jenkins, "we are still figuring out the magnitude" of these changes.

In the contemporary social movements described in previous chapters, the media they used, the networks they developed, and the messages they conveyed were all intertwined. Advocacy groups used communications technologies in ways that fostered trust and developed personal bonds by conveying inclusive, positive messages. The West Virginia teachers and the marriage equality movement in Maine harnessed social media to connect the experiences of different individuals the campaigns came across in their outreach. Organizers used Facebook and Twitter to find supporters, or, as in the case in Maine, to spread compelling stories of how individuals otherwise unconnected to the campaign were beginning to support the Maine referendum effort. By utilizing advances in technology, the organizers in West Virginia and Maine put a human face on their respective campaigns, campaigns that connected real people to the grassroots effort in authentic ways. In both cases, the outreach did not seem prepackaged or polished; rather, the sentiments that individuals shared were the spontaneous and heartfelt expressions of coworkers who were looking to help each other achieve the dignity in the workplace they felt they deserved or of neighbors and friends who had expressed genuine sentiments in favor of marriage equality.

The VAWA campaign also used this strategy to harness social media to spread the word about the effort to expand VAWA's protections. They enlisted a grassroots network of organizations and had their members share personal stories about what an expanded VAWA would mean to the average person. By using social media tools to spread personalized stories, the advocates created connections to the campaign on a personal and human level, again, putting a human face on the effort.

Perhaps a bit more pedestrian, the VAWA and marriage equality campaigns also used relatively modern means to stay connected to and help coordinate the grassroots efforts, like teleconferencing, e-mail, and other tools. The VAWA leadership network in particular used these means to foster effective communication, build trust, and strengthen alliances. Essential to these efforts was dialogue that created a personal bond among the leaders, whether it was formed through face-to-face meetings or by teleconference, with or without video. Ultimately, the strength of these alliances and the depth of trust that had been formed among these groups helped them weather the resistance they faced from skeptical and even hostile legislators.

The Long Beach campaign, although smaller in scope than the Maine or VAWA efforts, utilized the technology that was best suited for the scale of that campaign. Effective use of targeted print advertising and social media helped spread a series of inclusive messages that were each tailored toward different segments of the community. In this way, technology helped give the different target audiences a way to feel connected to the campaign, by highlighting the ways workers would benefit from the referendum, how small businesses would grow, or how the community as a whole would improve.

In these ways, contemporary groups are using different communications technologies to create personalized connections, whether those connections are inside the campaign—as in the case of VAWA's alliance-building efforts—or outside, as all four campaigns did effectively. Thus one of the first lessons of these contemporary campaigns—one national, two statewide, and one local—is that different media can be used in ways that create personalized connections that foster trust, stress interest convergence, and build a broad base of support.

Whether it is the split-second technology of social media, or the real-time, group communications made possible by conference calling, these modern technologies can help ground a campaign in personal connections, the type that build social capital, trust, and cooperation; help groups to coordinate their actions; and lower social distance between members of the network. These components seem essential for a movement for social change to be effective. Campaigns that can harness modern communications technologies to build social capital, as the four contemporary campaigns seem to have done well, can tap into the components of the social change matrix. This takes us back to the critique of the mass-mailing technology that seemed to take hold in the last decades of the twentieth century. Many national advocacy organizations used this technology to create powerful organizations that have had great impacts on law and policy over the course of the last forty years. Environmental organizations helped bring about sweeping legislative changes in the 1970s, like the Clean Air Act and the Clean Water Act. LGBTQ advocacy organizations promoted effective antidiscrimination legislation in many states. Advocacy organizations for women's rights have promoted a greater understanding of sexism and have made great strides in promoting gender equality. Depending on one's political views, one might think such efforts have had great success, have a long way to go, or have gone too far. My point here is not to debate the relative merits of these causes, although my political views, I would assume, are pretty apparent by now. The goal is to assess the relative success of these movements in bringing about the changes they seek.

It is safe to say that the groups that arose in the wake of the successes of the civil rights movement—environmentalists, LGBTQ rights activists, and modern women's groups, to name a few—had some degree of success in the late twentieth century. What we see from the successes of the contemporary campaigns described in these pages, movements that utilize technology to rebuild translocal, diverse networks, are those that may prove effective—perhaps more effective than the more centralized groups—in advancing social change, particularly in social innovation moments. While such efforts may be more effective on a smaller scale, as in the Long Beach campaign, even the national effort around expanding VAWA's protections had elements of a translocal campaign. Effective use of modern communications technology helped build and sustain that campaign in ways that created social capital among the leadership of the group. The trust that social capital developed did, in turn, filter down throughout the networked constituents those leaders represented.

Today, leaders have at their disposal new tools that can both create new opportunities for face-to-face interactions but also build on existing ones, harnessing the power of the strength of weak ties and creating network effects. Indeed, many of the national, centralized groups are teaming up with networks of grassroots groups and forming new networks, as we saw in the case of the marriage equality effort and as embodied in such entities as Faith in Action, the Poor People's Campaign, and the work of groups such as the Opportunity Agenda and the Center for Popular Democracy (more on this last group in a moment). National groups are partnering with local organizations and harnessing the power of technology to strengthen face-to-face, networked organizing.

To borrow the phrase from Brynjolfsson and McAfee again, speaking about workers in the workplaces of the future, leaders are learning to race *with* machines, not against them, and to leverage the power of digital tools to spur organizing and action at the local level.[6] Indeed, Facebook and other social media outlets can help leverage existing ties and spur positive social behavior, building on face-to-face relationships. Recall the study of voting behavior referenced in chapter 5.[7] Close friendships, those likely built on face-to-face relationships, when boosted by social media capacities, make a powerful combination to turn information into action.

The contemporary campaigns highlighted here have used modern communications technologies in effective ways that have not just built on existing relationships but also created personal connections and developed social capital. Organizers looking to advance social change will aim to harness the best available communications technologies in ways that build personal connections, foster trust, and create translocal networks of cross-class groups. The effective use of social media is one way to do this. Groups like the It Gets Better Project have done a remarkable job of persuasion through personalization. Mostly directed at LGBTQ youth, many of whom experience homophobia and alienation, the campaign attempts to connect adults who have made it through the crucible of the teen years and into adulthood and have found acceptance and support at the end of that journey. These independent spokespeople who participate in the campaign send a message that is fundamentally one of hope, but it is also one in which individuals, from artists and actors and musicians to President Obama, try to communicate on a personal level. They seek to inspire youth to persevere through any troubles they may be experiencing with the promise that they will find acceptance in the adult world. That personalized promise is designed to give them strength. These messages are posted on the organization's website and promoted through all manner of social media, from YouTube and Facebook, to Twitter and Instagram. Leaders and grassroots advocates can deploy the new tools of advocacy to connect individuals and organizations in ways like never before, and these tools can, in turn, help lower social distance, build personal connections, and increase trust. But these technological tools are means to an end, not an end in themselves.

We see in the examples in previous chapters that modern technologies—the video and telephone conference, e-mail—can help groups organize and communicate, build trust, bridge across organizations to link networks and increase network effects, lower the costs of organizing, and coordinate messages and actions for maximum social impact. In these ways, in addition to assisting a network communicate a personalizing and humanizing message, technology also helps groups organize, rally, and coordinate their actions. Thus technology is at the heart of not just a group's message, but also the network through which that organizing occurs.

Successful social movements in social innovation moments have utilized advances in communications technologies to do three things. First, they have used such technology to spread their message and build connections between leaders, organizers, and members. Second, that message has, more often than not, been a positive one, one that stressed shared humanity and was personalized to build on and strengthen social bonds and create an environment in which social capital could thrive. Effective use of medium and message requires a third component, however. Over the course of U.S. history, social movements that have had success in social innovation moments have tended to build translocal, diverse, grassroots networks that have often been shaped by the best communications technology available to them. What is more, the networks often looked like the medium they used to communicate. At least until the late 1960s, social movements that harnessed the best available communications technology of their respective times built translocal networks and were able to have success on a national scale, even though they were made up of hundreds, thousands, and even, like the American Legion, over ten thousand small "cells" of organizations. Although the radio and television were technologies that these groups put to effective use, they did not have unfettered access to these new means of communication. Compare that to the groups that would emerge soon after the victories of the civil rights movement; these organizations could engage in targeted, direct-mail marketing that was under their complete control. In the late 1960s and early 1970s, just as a range of new, progressive advocacy organizations were forming, encouraged by the successes of groups like the NAACP, a new technology was also taking hold, and this technology permitted such groups to forego the painstaking work of building translocal, cross-class networks. Instead, many of these groups often chose a different path. Empowered by the ability to create computerized databases of members from across the country, they formed national organizations that were, for all intents and purposes, separated from a true grassroots base. They also divided the electorate into specialized affinity groups, and, to a certain extent, may have catered to wealthier individuals (i.e., those who could make larger donations).[8] They thus lost out on the opportunity to build more diverse organizations. Just as many of the groups that formed in the Progressive Era were segregated by race, the new organizations that formed in the post-civil rights era were too often narrowly cast. While they failed to build translocal networks, they also failed, for the most part, to build more diverse movements, particularly those that had cross-class elements to them.[9]

In the contemporary era, the groups highlighted here are seeking to replicate opportunities that encourage face-to-face communication, creating personal bonds, building trust, and reducing social distance. Utilizing the best available communications technology, whether it is social media or simple tools like the ability to engage in conference calls, groups are harnessing such technology to create personalized connections, expanding opportunities for face-to-face communications, and building social distance-reducing bonds. They are also seeking out opportunities for advocacy on a scale that gives rise to such tactics, as in the Long Beach campaign, described in chapter 8.

In that effort, which was launched only on a citywide basis, the coalition formed to pursue an increase in the minimum wage for workers in larger hotels. It utilized a range of technologies that helped build personal connections, leveraged existing social capital, and emphasized interest convergences. The coalition was, indeed, a network: a network of several organizations at its hub, but with a diverse range of affiliated groups, from homeowner associations to environmental groups. And they harnessed technology to spread their message and keep tabs on the alliances they were building. Because the campaign was citywide in scope, the need to deploy social media was low, though the coalition certainly utilized it. Instead, they used computerized systems for keeping track of supporters and potential allies, but such systems were no substitute for the hard, face-to-face work necessary to move the campaign forward. They coordinated their mobilization efforts and matched volunteers and staff with the groups with which they were affiliated, harnessing their existing social capital to build a loose network of "strong-yet-weak" ties that would support the ballot measure. At the heart of this approach, which mimicked the fractal nature of a nested enterprise, were the personal connections the coalition attempted to utilize and leverage at the "ends" of the network. Individuals at those ends were encouraged to speak to each other and were not required to funnel communication through a centralized hub. Indeed, an active participant in coalition efforts would be matched with his or her neighbors and friends in other civic groups of which he or she was a part. Thus the coalition took advantage of existing social capital and the resulting bonds of trust to bring in new supporters to the effort. Technology helped track these connections and allowed the coalition to activate a decentralized, engaged, and diverse network of supporters once it came time to vote on the referendum. In these ways, the technology fostered face-to-face communications that built on preexisting trust and capitalized on interest convergences. It also created personalized connections between coalition supporters.

Technology may have played a small—but critical—role in the Long Beach campaign, but its importance appears to grow when the scale of an effort increases. In the Maine marriage equality campaign and the efforts of the West Virginia teachers, both statewide efforts, the organizers there built movements based on face-to-face organizing that was coordinated by and utilized modern forms of technology. Computerized systems helped keep track of supporters in Maine, while social media helped spread the word about both campaigns, in messages that humanized the movements by highlighting the personal stories of support. In Maine, canvassers collected these stories by going door-to-door and having face-to-face communications. In West Virginia, the teachers found each other in the staff rooms and over social media and came together in rallies and while organizing meetings for face-to-face communication made possible by and coordinated through digital tools. Social media thus became a way to echo and amplify the stories the canvassers and organizers were discovering in their outreach. In Maine, in-depth, face-to-face communications with potential voters in the field unlocked powerful stories of why average voters were willing to support the marriage equality vote. In both settings, a fractal array of organizations leveraged modern technologies to create a network of supporters, all the way down to the "grassiest" of grassroots. Both networks harnessed social media to amplify their messages and digital tools to find each other, coordinate outreach efforts, and keep track of supporters. In Maine, statewide and national groups stayed in contact and communicated about the campaign through conference calling and e-mails. Technology enabled the leadership to receive instantaneous reports from the field so it could remain in constant contact and adjust the messaging and outreach efforts in real time.

While national campaigns may not have such direct grassroots reach, by coordinating networks of groups—which is what true, translocal organizations always have been in practice—coalitions of organizations can build far-reaching support and use technology to do so. In the VAWA reauthorization campaign, the National Task Force to End Sexual and Domestic Violence Against Women was a coalition of groups that had grassroots reach. Leadership of a range of organizations—from women's groups, statewide coalitions providing services to survivors of intimate partner violence, LGBTQ organizing groups, Latina-serving organizations—came together to form the steering group of the task force network. These leaders became the representatives of their respective constituents and served to move the work of the coalition forward. They represented their respective constituents' interests and served as a communications relay between the work of the network and the members of the organizations that made up that network. Technology aided these activities as conference calling and e-mail helped coordinate these efforts. At the same time, face-to-face communications proved necessary to sustain the campaign and helped build bridges and foster trust among the leadership of the network. This trust then filtered down throughout the network, into the organizations those leaders represented. Modern technology helped the leaders stay connected, though, to communicate about activities in real time and at low cost, give reports on lobbying efforts, share information learned from the field, dispel rumors, and spread news of victories. The network also utilized social media to stay engaged with the members of various coalitions involved in the effort. By using clever messaging, they disseminated information down and throughout the network and unleashed the grassroots to spread information back up through the network and beyond. This effective use of social media echoes its use in the West Virginia teachers' campaign and Maine marriage equality effort and helps show how social media can spread messages, encourage participation, highlight personal connections, and foster engagement. Leaders can then convert such engagement into broader participation when it comes time to vote, contact an elected official, or encourage a friend to get involved in the effort. The VAWA reauthorization movement also formed into a fractal network that was connected by technology. This national network ultimately had grassroots reach as the steering committee's members represented coalitions, other networks, and groups that gave the network deep and widespread reach within the advocacy community. All of this was facilitated and made possible by communications technology that helped this network function effectively, and in ways that spread trust: building on existing social capital (the groups that formed the coalition) and creating new social capital (as leadership of these coalitions were able to form bonds with leaders of other organizations). This is bridging social capital at its best.

What these recent campaigns also show is that "cross-interest," or what the legal scholar Kimberlé Crenshaw might call "intersectional," organizing is also critical to social movement success.[10] In the Long Beach campaign it was not only environmental groups, but also groups of homeowners, that were asked to support the effort, through a campaign that stressed interest convergence. Similarly, in the VAWA campaign, more established and mainstream women's rights groups paired with LGBTQ organizations to create a stronger and more united front, one that could leverage the strengths of all of the groups to achieve the goals sought by each organization within the network. These examples show that contemporary social movements appear to be moving away from some of the strict, siloed, identity-based politics of the last forty years. Groups that formed in the 1970s were not just national organizations without a grassroots base; they were often focused on a particular identity-based group. Groups that originally formed as the Puerto Rican Legal Defense & Education Fund and Lambda Legal Defense & Education Fund got their start during this period and focused on more identity-based efforts. Perhaps what we are seeing in the examples of the more contemporary campaigns is that identity-based organizing might have its limits, and that cross-issue, cross-interest, and "cross-identity" efforts are needed to advance more robust social change.

There is, of course, a place for identity-based advocacy. If this book says anything, it is that groups must form in structures that produce trust, and there is certainly a need for individuals, particularly those from marginalized and oppressed groups, to have a place where they can speak and think and act in an environment that is safe and conducive to trust. Many may feel that such identity-based spaces have to exist in order to foster those feelings of trust and mutual support. I agree. Sunstein argues that such "enclave deliberation" is also necessary for experiments with social change, as groups test out how to confront existing norms and develop new norms that can spread to other groups.[11] This makes great sense. At the same time, I do not ascribe the growth of identity-based organizations to an embrace of individualism ushered in by the Reagan era, as political scientist Mark Lilla does.[12] He argues: "For the fascination, and then obsession, with identity did not challenge the fundamental principle of Reaganism. It reinforced that principle: individualism."[13] But the growth of identity-based groups in the wake of the successes of the civil rights movement, while Reagan was governor of California and before anyone had ever coined or uttered the term "Reaganism," was likely a function of communities identifying ways in which their members were similar, seeking to bond over injustice, spark interest in social change, and cultivate spaces where individuals could come together for mutual support. But those spaces must be linked to other spaces, where individuals of different backgrounds can come together, engage in communication, lower social distance, seek to support one another, find common ground, activate weak ties, build network effects, and support larger movements for social change. Ideally, new spaces can form around which individuals of different backgrounds can come together, discuss pressing issues, seek commonality and shared interests, and move forward together.

Such spaces can only arise and exist if there is both trust at the outset, and those spaces, themselves, generate trust moving forward. Bridging social capital can help with creating that trust at the outset, the idea that groups will connect with each other through common links. When I was a young legal aid lawyer working in Harlem and Washington Heights in New York City, predominantly representing low-income tenant associations made up almost exclusively of African American and Latinx tenants, my capacity for success in advocating on behalf of those groups, the only way I got my foot in the door, was a result of the trust perhaps one or two members of those groups had in my organization, the Legal Aid Society of New York. Perhaps someone in the group had a good interaction with a legal aid lawyer like me in the past. Sometimes an individual in a neighboring building had talked about the role of my organization in helping tenant associations obtain desperately needed repairs or prevent a landlord from charging illegal rents.

Individual members of the tenant associations would have to convince their members to at least give my organization a chance and invite a lawyer into their homes to discuss the potential for working together. These individuals were connectors. They had a belief that my organization might help the group and they leveraged their own social capital to make the connection. This happened in Long Beach as well, as members of intersecting groups built on preexisting relationships of trust to advocate for an increase in the minimum wage for workers at large hotels in the city. Because of such preexisting relationships, individuals who were supportive of the cause were able to connect the campaign to other groups that revolved, like satellites, around it. Leveraging their own social capital and their relationships of trust with the members of those groups, they were able to bring that new group into the gravitational pull of the campaign. Building trust at the outset, getting in the door, is a product of bridging social capital. Members of groups must connect their seemingly unconnected groups together, serving as the glue between them. This is what some Long Beach citizens were able to do: build on their own preexisting relationships to connect the minimum wage campaign to organizations like homeowners' associations and environmental groups, neither of which might be seen as likely allies in a campaign on behalf of low-wage workers.

What is more, developing "trust moving forward" is a product of active civic groups. Trust begets trust in a virtuous cycle.[14] Trusting and trustworthy behavior develops and grows where individuals know they will have opportunities for future interactions, where social distance is lower, and where they engage in communications generally, but also when such communications involve agreements to cooperate. Individuals can lower social distance by finding what they share as opposed to identifying their differences. All of these opportunities for developing trust arise in small-group settings where individuals can discuss their commonality and shared goals and agree to work together moving forward.

Just as social capital theory recognizes that there are two types of social capital, bonding and bridging, perhaps there are two types of social organizations and two types of social movements: those that create bonding between individuals with particular backgrounds, and those that help create bonds across backgrounds, building on the strengths of each to create a whole that is greater than the sum of the parts, which is the essence of network effects. Close-knit organizations and movements that help bonding social capital thrive help people "get by." At the same time, bridging spaces and bridging movements advance a cause in the broader society and help individuals and groups "get ahead." In what Castells calls the "network society," the creation of the network allows for elements of the networks to produce and achieve more than they might otherwise on their own.[15]

When a movement attempts to get ahead it tends to leverage the strengths of bonding groups, across identities. It uses an interest-convergence approach to show that it is in the interests of a broad cross-section of organizations to strive for a particular outcome. We saw this in the VAWA reauthorization effort and the campaign for a living wage in Long Beach as seemingly disparate groups came together to advance a particular outcome, one that held out the possibility of wide-ranging benefits. In the VAWA campaign, a renewed VAWA was in the interests of the original groups that had campaigned for VAWA over the years. At the same time, a strengthened VAWA, with broader protections, attracted a new set of groups into the advocacy fold. Similarly, in Long Beach, an organization that traditionally advocated for low-wage workers, a union, came together with groups that do not usually team up with unions, like small business owners and homeowners' associations, in an effort that had both wide support and broad-ranging benefits. When bonding organizations come together through bridging efforts, these campaigns seem to harness the power of network effects, offering a greater chance of success.

In the marriage equality campaign, while LGBTQ groups were at the forefront of what has turned out to be a monumentally successful national effort, heterosexual allies and groups that were not typically focused on LGBTQ issues supported them. For example, shortly after President Obama came out in favor of marriage equality in 2012, the NAACP passed a resolution professing its own support for the effort. For the NAACP, this support was tied to its antidiscrimination work and its belief that the U.S. Constitution's Fourteenth Amendment protects this fundamental right for the LGBTQ community.[16]

Bringing bonding groups together through bridging networks appears essential to movement success. Organizations that made up the civil rights movement of the 1950s and 1960s were often cross-class and cross-race, and sometimes civil rights leaders took pains to recruit across such differences in the belief that it would strengthen their efforts.[17] Think of the Freedom Riders, a collection of cross-race and cross-class volunteers who worked in the South to protest segregation and register African Americans to vote.[18] At the same time, many organizations in the Progressive Era had cross-class elements to them. Unlike the NAACP that formed at the tail end of that era, however, few of the organizations from that period were integrated along racial lines. As a result, perhaps, the state of civil rights for African American citizens at the time deteriorated, particularly in the South.

Is the lack of progress on African American civil rights during the Progressive Era, when the Jim Crow system mostly came into being, a function of the fact that many of the Progressive Era organizations were not racially integrated? Strict identity-based organizations may offer few opportunities for lowering social distance across identity-based lines. Without that lowering of social distance, without the opportunity to share perspectives and build trust, advancing the cause of African American civil rights during the Progressive Era faced significant challenges, despite heroic efforts of grassroots organizers like Ida B. Wells. We have seen in the past that the technology available for organization building can have an impact on social movement structure. And social movement structure, in turn, can have an impact on the extent to which that movement helps generate trust and lower social distance. Perhaps, in addition, the extent to which an organization works to bridge different divides—race, class, gender, sexual orientation—also determines the extent to which that movement is bonding or bridging in character. And the difference between whether a group is bonding or bridging may also determine its potential for success, particularly when a group tends to be homogenous in a particular way.

The modern period, as I will call it, which was marked by national organizations that did not consist of translocal and cross-class groups, is also the period in which economic inequality began to grow. From the Great Depression through the early 1970s, economic inequality was in check and diminishing. As civic participation thrived, a strong economy helped usher in a roughly four-decade period in which rising wages, higher tax rates, expanded participation in higher education (spurred by the G.I. Bill), and widespread prosperity reduced economic inequality considerably. The first three decades of this period also saw the beginning of the modern civil rights movement, and its successes in the 1950s and 1960s helped reduce the economic inequality created by Jim Crow and other forms of race-based political and economic inequality.

As described in chapter 4, Tocqueville observed that no aspect of American life in the early nineteenth century "struck [his] eye more vividly than the equality of conditions."[19] This would prompt him to ask whether there was a "necessary relation between associations and equality."[20] Two hundred years later, to this observer, his question still resonates.

When an association is a product of bridging social capital as opposed to one based on bonding social capital alone, there is the greater likelihood that such associations will have a greater effect on inequality as between members of the network. In such associations, trust is both a precondition and a result of effective bridging social capital. Bridging groups, because they allow for give-and-take between members and across social distance, tend to embrace more inclusive messages that are crafted from the dialogue that arises in diverse groups. In fact, cross-cutting messages are more inclusive and can have the added value that they invite a broader cross-section of organizations, with members from different backgrounds, to join the effort. The contemporary use of the hashtag "#MeToo" is perhaps a paradigmatic example of the cross-sectional, bridge-building, and personalizing capacity of message-making in the Internet age. What the history of at least some of the social movements that have emerged in social innovation moments seems to suggest is that bridging social movements not only organize themselves in inclusive ways, they also embrace more inclusive messages built on ways in which the members' interests converge. And it is because of this combination that they, in the end, tend to produce more far-reaching results, particularly in terms of smoothing out the differences between their members, as Tocqueville recognized. The social science literature sometimes distinguishes between what are called consensus movements, those that can enjoy as much as 80 or 90 percent public approval, and conflict movements, those that attempt to change the fundamental social structure of society and do not enjoy such broad support.[21] According to Bell's interest-convergence theory, however, significant social change requires one to not just seek such broad change, but also to attract a broad, cross-section of society.

Saul Alinsky, referenced briefly earlier, and a common target of conservative conspiracy theorists, actually was not as purely radical or outside the mainstream as one might typically think, and he believed that different groups needed to come together to enact positive change. Sometimes associated with radical elements in popular culture, Alinsky was actually quite tempered in his views, particularly when it came to building coalitions, often taking a very pragmatic approach. For example, he would proclaim (admittedly, using the gendered language of his time) that if a long-haired community organizer finds that "having long hair sets up psychological barriers to communication and organization, he cuts his hair."[22] In similarly dated language, he stressed the importance of interest convergence across different demographic groups in a call to create a groundswell of popular support around policies attractive to a large cross-section of society: "We are belatedly beginning to understand this, to know that even if all the low-income parts of our population were organized—all the blacks, Mexican-Americans, Puerto Ricans, Appalachian poor whites—if through some genius of organization they were all united in a coalition, it would not be powerful enough to get significant, basic, needed changes. It would have to do what all minority organizations, small nations, labor unions, political parties or anything small, must do—seek out allies. The pragmatics of power will not allow any alternative."[23]

Alinsky would criticize what he saw as organizations that had too narrow a focus, where they worked on just one or two issues. "In a multiple-issue organization," he would write, "each person is saying to the other, 'I can't get what I want alone and neither can you. Let's make a deal: I'll support you for what you want and you support me for what I want.' Those deals become the program."[24] For Alinsky, a narrowly focused group that does not seek such coalitions will have little staying power: "Not only does a single- or even a dual-issue organization condemn you to a small organization, it is axiomatic that a single-issue organization won't last. An organization needs action as an individual needs oxygen. . . . Multiple issues mean constant action and life."[25]

Grassroots organizations that might have a narrow focus for their immediate advocacy must seek out coalitions with other, possibly narrowly focused groups, and get them to support each other in their respective efforts. In an artful campaign that echoed Medgar Evers's speech about being a Mississippian, the Long Beach organizers stressed the fact that the low-wage workers who would benefit from an increase in their wages were Long Beach residents. They were from households that would spend their increased income in small businesses in Long Beach. The shared personal characteristic (local residency) coupled with a shared interest (increased local economic activity) helped build a broad-based coalition centered on not just interest convergence, but also on sharing an essential personal trait, the recognition of which helped to lower social distance.

In addition to stressing the shared trait of local residency, the Long Beach coalition also took great pains to build personal connections in support of the campaign. This was done through connecting messengers from similar demographic groups as the specific target audience and tapping into previous social capital connections. Knowing that a civic group, like a block or homeowners' association, would likely be more receptive to overtures from a member, supporters of the campaign were tapped to reach out to their preexisting social networks to garner support for the effort. The former labor organizer Jane McAlevey calls this work "whole worker organizing": situating a worker in her broader community and leveraging those broad relationships to advance a group's cause. In Long Beach, campaign volunteers from different backgrounds were asked to connect with those with whom they shared demographic traits.[26] By building on these sorts of instances where social distance might already be low, the campaign was able to garner broad support for the living wage referendum. The organizers of the Long Beach campaign also looked beyond the face-to-face interactions to build personal connections, connections that would either take advantage of already low social distance, or were designed to personalize and reduce that social distance, creating a personal bond with the campaign. In its marketing materials, the campaign used photographs of Long Beach workers and small business owners, from different demographic groups, in an attempt to build connections between likely voters and the campaign. The hope was that by putting a friendly and familiar "face" on the campaign, potential supporters could develop a personal connection to the effort. Similarly, the door-to-door canvassing work to garner the signatures necessary to place the wage measure on the ballot was designed around trying to have in-depth conversations about the campaign in the hope that, once again, potential supporters would be able to put a face on the campaign, and, when it came time to vote, they would have developed a personal connection to the effort.

As we have seen in recent campaigns highlighted here, specifically the marriage equality movement and the effort to strengthen VAWA, positive, inclusive messaging is also key to tapping into the social change matrix. For the marriage equality advocates, getting away from terms like "gay marriage" or "same-sex marriage" was part of an effort to broaden popular support for the campaign. Similarly, in order to both renew VAWA and obtain enhancements to its protections, changing the way organizers would speak about what has been traditionally referred to as "domestic violence" in favor of the more inclusive "intimate partner violence" helped build broad-based support for efforts to extend and strengthen the law.

Recent research helps show the strength of these sorts of broad-based messages, particularly across racial and economic divides. Researchers canvassed eight hundred homes in Minnesota and showed residents a flyer that offered what they called "classic dog whistling," calling for tax cuts and scapegoating immigrants. Then half of the white respondents were shown a flyer with a traditional Democratic message that was silent on race, while the other half were shown one that stated the following: "Whether white, black, or brown, 5th generation or newcomer, we all want to build a better future for our children. My opponent says some families have value, while others don't count. He wants to pit us against each other in order to gain power for himself and kickbacks for his donors." When showed the message that was silent on race, a majority of white voters chose the conservative candidate. When showed what the researchers called the "race-class" message, 57 percent of the respondents would switch their vote, leading the researchers to conclude as follows: "The race-class message was significantly more effective at generating progressive votes than the class-only script. And this was among the white voters initially keen on the divisive message."[27]

One of the components of social movement messaging, which is consistent with the interest-convergence thesis, is that social movements, in order to build trust and find those areas of common ground connecting otherwise disparate groups, seem most effective when they are delivering messages that lower social distance between the messenger and the party receiving the message, which is often a product of personalized communications that stress how individuals and groups share common interests and a common humanity. Several years ago, the Center for Popular Democracy (CPD), a national organization that attempts to connect policy advocacy with grassroots groups from across the country, embarked on a campaign to help shape the fiscal policies of the Federal Reserve Board. The Federal Reserve often operates literally behind closed doors, and the leaders of this campaign, which was entitled "Fed Up," worked to expose Federal Reserve leaders to the real-world faces and stories of individuals on the low end of the income scale harmed by national fiscal policy. They did this by having individual low-wage workers appear outside of meetings where Federal Reserve leaders were convening so that they might confront them in person and in a face-to-face fashion. This type of advocacy helped the bureaucrats put a human face on the fiscal policies they were adopting, and the leadership of the Fed Up network claims credit for softening some of the Federal Reserve's policies and advocating for greater diversity among the ranks of the officials that make up the government body.[28] What one of these leaders would ultimately find, however, is that this sort of personalized and humanizing advocacy would take an entirely different tone and salience when, in some ways, he became a part of the story.

Ady Barkan was a young, aspiring journalist when he decided he wanted to go to law school. He believed a law degree might enable him to have more of a direct impact through his work. He also thought he might want to run for political office someday because he "loved giving speeches and engaging in debate."[29] After law school, he went to work first at the Brooklyn-based grassroots group Make the Road, which advocates for low-wage workers and tenants out of its main office in Bushwick, a low-income community in New York City facing the impact of gentrification. He ultimately joined the CPD staff and helped launch the Fed Up campaign despite some skepticism about whether a grassroots effort could penetrate the policies of this opaque bureaucracy. Several years of tireless advocacy building grassroots support for the effort, creating a local network that would help to confront Federal Reserve officials, raised the profile of national fiscal policy such that it became a source of debate and discussion during the 2016 presidential election. But late in that election cycle, in October of 2016, Barkan and his family got a devastating diagnosis: he was stricken with the terribly debilitating amyotrophic lateral sclerosis, or ALS, often referred to as Lou Gehrig's disease.

Barkan's work took on a whole new dimension as his diagnosis put him front and center of the health care and tax policy debates that arose in the first year of the Trump presidency. While he was never the shy and retiring type, his previous work was consistently done on behalf of others. Now he was becoming the centerpiece of his advocacy. As he explains, despite the fact that he "never was a kind of behind-the-scenes guy," and he always thought of himself "as a protagonist in a political story," prior to his diagnosis, this work was "done in a kind of 'on behalf of others and the world that I want to create approach' as opposed to 'I'm the impacted person' type of story." Now, because of his illness, he would find himself affected directly and personally by some of the social policies promoted by Republicans in Congress. He set out to oppose such policies partly because of his political leanings, but also because these policies would, he feared, have an adverse effect on his health and the ability to obtain the health care he would need in the immediate future.

Some of these issues came to a head because of a fortuitous and chance meeting Barkan had with the Arizona Republican senator Jeff Flake. As fate would have it, the two found themselves on a commercial flight together, and Barkan was able to convince Senator Flake to discuss his pending vote on the massive Republican tax cut. Fearing that the cut would lead inexorably to higher deficits and calls for cutting back entitlements and health care spending, Barkan explained that such cuts would jeopardize his health care and ultimately his life. Barkan's face-to-face conversation with Senator Flake on the plane was recorded on video and immediately went viral. In labored speech, a product of his illness, Barkan asked Flake to think about the senator's legacy, Barkan's infant child, and Flake's grandchildren. Barkan also encouraged Flake to turn his principles into votes against the tax bill. He added: "My life depends upon it."[30]

Later, taking a page from the Fed Up playbook, Barkan and others flocked to the nation's capital to take on the tax bill, many wearing T-shirts with the hashtag "#listentoourstories" that reflected the approach of the group: confront Republicans in the Senate in particular with personal stories of the impact that the tax bill would have on the lives and well-being of the elected officials' constituents, an effort to lower the social distance between them and help the politicians see the impact of their polices on real people.

Barkan is not one to put a positive spin on his illness, however. He does not say his illness is the best thing that ever happened to him. Quite the contrary. He would "happily trade it in because I very much like doing the work I was doing."[31] Nevertheless, he brings the fearlessness that animated his Fed Up work to his advocacy to address some of the issues he faces as a consumer of health care. He is considering what he can do effectively in the time he has remaining before his illness completely saps him of his ability to serve others. And he is proving the power of social media amplify these stories, build personal connections, lower social distance, energize advocates, and help people see the connection between their own lives and national policies.

For Barkan, social media really gives people tools to amplify their messages quickly and to organize for collective action. He describes the aftermath of the Valentine's Day school shootings in Parkland, Florida, and the groundswell of advocacy that emerged as follows: "The social media facilitated locally rooted national organizing and getting people into the streets; into conversations with one another in their communities; targeting local political actors, giving money, registering voters, as a result of some speeches that went viral on social media. And then the mainstream media picked it up."[32] At the same time, Barkan is concerned about "thin organizing," which can happen with social media. With such organizing, the action of just a small number of individuals can catch fire and capture the imagination of the social media universe, but it does not have the same sort of power that more sustained, deeper organizing can produce. For Barkan, "the real sweet spot is when you can combine real organizing with good social media work."[33]

The Downsides of Digital

Finding this sweet spot is essential to social movement success in the present social innovation moment. At the same time, one of the things that anyone engaged in organizing in this current social innovation moment—with all the capacities that contemporary technologies afford—will also have to grapple with is the potential and serious downsides these technologies present. The fact that digital tools are so easy to use means they can sap movements of their participants, not by helping such movements overcome the free rider problem, but by making, if not "free" riders, perhaps "fake" riders. This problem is sometimes referred to as "slacktivism," or, as Lilla describes it, the "Facebook model of political engagement":[34] the notion that individuals may engage with a movement by liking a Facebook post or forwarding a tweet with a hashtag in it and feel they have done their part for the movement.[35] The fake rider is perhaps more pernicious than the free rider, because that individual feels like he or she is actually supporting the group's effort, when that minimal input will not in fact lead to any real change and is unlikely to lead to the sorts of benefits of social capital formation we have described here.[36] The phenomenon led Malcolm Gladwell to rail against this sort of activism in an article for *The New Yorker*, the subtitle of which declared that "the revolution will not be tweeted."[37] Some saw a response to Gladwell coming in the form of the social media-fueled activism of the Arab Spring.[38] Still, these are very real concerns. Similarly, the ease with which one can use these tools to generate and activate popular action can also lead to antidemocratic activities, like the torch-lit alt-right rally in Charlottesville in 2017, or the fake, conflicting rallies hosted by fictitious identity-based groups in the lead-up to the election of 2016 supposedly engineered by individuals manipulating social media to create chaos and conflict and turn people away from the political process.[39]

Are digital organizing tools more prone to these sorts of problem? Definitely. Are the analog mechanisms available to groups to monitor the free rider—face-to-face engagement, communication that helps lower social distance, repeat play—available to rein in the fake rider or the fake organizer as well? Definitely, again. Digital tools are effective mechanisms for outreach and coordination, but they must lead to the sort of efforts that yield the more traditional forms of social capital: face-to-face meetings, joint activities, marches, rallies, and so forth. Digital tools have a significant role to play in overcoming the free rider problem by making participation and coordination easy; the downside of these tools requires more traditional mechanisms to ensure they can lead to the sort of engaged advocacy required of sustained collective action.

Similarly, the decentralized nature of communications over digital networks means that so-called fake news can flourish and spread, and often at a much higher rate and faster speed than accurate information.[40] To quote the journalist Franklin Foer: "Facebook and Google have created a world where the old boundaries between fact and falsehood have eroded, where misinformation spreads virally."[41] But the filtering process that can occur through groups communicating about such news in a decentralized way, either digitally or through their face-to-face contacts, can help overcome the fake news problem. Writing on the precipice of World War II, the Hungarian-born sociologist Karl Mannheim feared (and saw) the impact of propaganda and believed the contemporary social structure of the time was helping it to flourish. He wrote as follows:

> The fundamental question now arises whether we can discover the origins of the rational as well as the irrational elements in modern society. Modern society, which in the course of its industrialization rationalizes larger and larger numbers of persons and ever more spheres of human life, crowds together great masses of people in huge urban centres. Now we know, thanks to a psychology absorbed in social problems, that life among the masses of a large town tends to make people much more subject to suggestion, uncontrolled outbursts of impulses and psychic regressions than those who are organically integrated and held firm in the smaller type of groups. Thus industrialized mass society tends to produce the most self-contradictory behavior not only in society but also in the personal life of the individual.[42]

One of the original responses to the "problem" of free speech—that anyone can say virtually anything—is that it is more free speech that can help overcome the falsity of such claims by discussion, deliberation, testing of ideas, and disabusing people of false concepts.[43] This process can be one that is a product of the trusting relationships one has through one's connections, both digital and analog, but it can also serve to build relationships of trust because, as we have seen, communication and dialogue can build trust, lower social distance, and create social capital.

Such trusting relationships can be harnessed to take another form of action as well: help combat fake news. By direct communications to each other, even if over social media, where we curate the information we receive, assess it for its truthfulness (probably based on our knowledge of the trustworthiness of the source), and only share that information which we consider legitimate with those we trust, we can help undermine the strength and spread of fake news. We tend to rely on information that comes from trusted sources, whether those sources are a particular news outlet or a close friend. The more individuals are conscious of assessing the source of the information to determine its trustworthiness, the more likely we are to be in a position to combat fake news.

One of the drawbacks of sites such as Facebook that use algorithms to curate the information users will see on their news feeds is that such algorithms can alter the decentralized nature of social media, changing the dynamic and the direct connection between users. Moreover, to the extent these algorithms read the reactions of one's contacts on these social media sites to the content provided and elevate certain items of information based on those reactions, this can create a sort of groupthink. As Cathy O'Neil describes in her 2016 book *Weapons of Math Destruction*, this runs the risk of "plac[ing] each of us into our own cozy political nook."[44] Are problems such as groupthink and confirmation bias more likely to play a role in communications over social media? Can such phenomena help *spread* fake news more quickly? Yes and yes. But just as bonding social capital can serve as a way of solidifying one's view, confirming a possibly distorted view of the world, bridging social capital may serve as another line of defense against fake news, and perhaps a stronger one than that which comes from closer bonds.

It is our bridging social capital, not our bonding relationships, that can not only prevent the spread of fake news, but also correct it. In a fascinating paper that tested what the authors called the "Emperor's Dilemma," named after the Hans Christian Andersen fable about a naked ruler, the researchers showed how unpopular norms and beliefs can spread in tight-knit groups that have few bridging contacts to other groups. At the same time, bridging contacts can break the hold of such norms and knowledge by bringing in outside perspectives that are not bound by the groupthink evident in the smaller group.[45] Similarly, and not surprisingly, in a recent study of the media ecosystem, with a particular focus on how it has operated in the last several years in the political sphere, researchers showed that media outlets that engaged in what they call a closed, "propaganda feedback loop" rather than engaging in an open, transparent, participatory, rigorous process of a news gathering and fact-checking governed by the norms of journalism tended to generate and spread more fake news stories.[46] The solution to these problems seems to be the type of approach that looks to harness the power of bridging ties rather than those reflective of closed, bonding relationships.

Apart from the fake rider and fake news problems, another significant issue, which has become much more salient recently, is the idea that the means of communication must be, itself, trusted and safe. The colonists faced this problem in the 1770s when they feared that the official postal system was not a secure means of communicating. The Continental Congress agreed to create its own system of transporting the mail and revolutionary tracts. In order for a means of communication to serve the ends of coordinating cooperative, collective action, the users of any system of communication must have confidence in the system itself, that it will not be used to undermine the collective effort, or, in some other ways, abuse the trust of those users. That is why today some prefer applications like WhatsApp and Signal, believing those networks to have better security and encryption. Social media and other contemporary tools of communication certainly hold out the promise of decentralization: that the means of communication, when placed in the hands of the communicators directly, will, for the first time, enable such users to control the message that a group's members can communicate. But such decentralization must not become a "Trojan horse," a way to lure people into a false sense of security so that their candid communications might someday face exposure. Under authoritarian rulers, groups are finding that the means of communication can be monitored, blocked, or used in ways to undermine any antiregime organizing that occurs. In the United States, organizers are beginning to realize that even in a nonauthoritarian regime, while the government might not yet utilize the means of communication to undermine organizing, there are two ways at least in which users can lose faith in the means of communication, or the "platform" in common parlance.

In the spring of 2018, it came to light that so-called private information of tens of millions of Facebook users was harvested by an individual who shared such information with a data firm aligned with the presidential campaign of Donald Trump. The breach, although it is not really appropriate to call it a breach because Facebook permitted access to this information, occurred back in 2015, when an individual purporting to conduct academic research created an app that asked users who wanted to take one of the thousands of so-called personality quizzes that are so ubiquitous on social media. (E.g., "Which *Downton Abbey* character are you?" or "Which slightly misremembered *Office Space* quote are you?" According to BuzzFeed, that last one is actually real.) When users took the quiz this researcher had created, they not only gave him access to their own information; he also got access to the private information of all of those individuals' Facebook friends, a treasure trove of as many as 87 million users' personal information and communications. In turn, this information was transferred to Cambridge Analytica, the data firm. It was then purportedly used by the firm to create personality profiles on these users, and that information was likely used by the Trump campaign in its messaging.[47]

This breach of trust has created a problem for Facebook and for social media companies generally. They like to give the product away for free to end users, but their business model requires that they not only offer ad space to advertisers who will pay for the privilege to have access to these sites, they also sell the data itself, so that those who want to use it to microtarget their messages to specific individuals to persuade them to take action: buy a product, vote, or, as the case may be, not vote. Indeed, one of the messages that was often shared in the lead-up to the 2016 election was not just voter motivation but also voter suppression—convincing members of certain communities to stay home on election day, as a group purporting to represent African Americans sent out a message that neither candidate was good for the African American community so individuals from such communities should not vote, assuming that most would vote for Hillary Clinton were they to make it to the polls.[48] Facebook founder Mark Zuckerberg has apologized to the community of Facebook users and recognized that the incident was "a major breach of trust."[49]

Just as in past experiences with advances in communications technologies, there are great risks when such technologies fall into, or are seized by, the wrong hands, precisely because of the capacities they hold for manipulation and abuse. The Nazis thought the radio could stir up sentiment and support for their campaigns, so their leaders thought it critical to ensure every German household had access to the "people's receiver"—an affordable radio—so that they might feed their messages into every home.[50] There are similar risks with the Internet and social media. Mobs from Charlottesville to Myanmar use social media to spread hate and violence. Their members are also able to find one another more easily. For McAfee and Brynjolfsson, the openness of the Internet has "led to great benefits for the world," but it has also "led directly to malware, cybercrime and cyberwarfare, distributed denial-of-service attacks, phishing and identity theft, darknets for exchanging child pornography, doxxing, fake news, and other developments that can make one despair for humanity."[51] They have simple advice for creators and managers of social media platforms though: "The solution to all this bad behavior and bad content is to build better platforms—ones that use curation, reputation systems, and other tools to keep out the bad and encourage the good, however the platform owner defines these terms."[52]

These problems with social media platforms and their relationship to social movements reveal yet another organizing paradox in the contemporary age: digital tools can spur action and coordinate cooperation, but they can also undermine end users' trust in the platforms themselves, which would lessen the powerful organizing potential these tools offer. While there are many suggestions for how to combat the trust problem that platforms face—from independent fact checkers to address potential fake news, to machine learning that can try to detect fake news by an assessment of the source—there is a very analog way in which consumers can advocate for social media platforms to become more trusted and trustworthy: organize. As the history of communications technologies described in the first three chapters reveals, the means of communication has often become a target of advocacy itself. The Continental Congress needed its own postal system. The competing voices on civil rights in the 1960s fought for airtime on the television and radio. Every advance in the means of communication seems to have led to some dispute over how that innovation would operate, using traditional organizing methods to do so. It should come as no surprise that social media should become not only a platform for communication, but a target of advocacy itself, to keep it accessible and trustworthy.

At present, it seems that technology companies are looking for a technology-based approach to curb the harmful effects of their products. But if they cannot find a tech-focused solution, they throw up their collective hands and say nothing can be done. What is worse, they likely think that there is nothing wrong if there is not a tech-based solution. There is an old saying in progressive lawyering circles: if it offends your sense of justice, there's a cause of action. In other words, if I think there is some injustice in the world, there must be a way to sue over it. But that approach has an unfortunate side. For some, if they cannot see their way to bring a lawsuit over some situation, they conclude there must not be anything wrong with it. Like the technologists, they throw up their hands and say nothing can be done. As with technology, the situation devolves into something worse: it leads one to conclude that the situation doesn't offend one's sense of justice if there is no lawsuit that can be filed to challenge it. In other words, to the lawyer, it can't be wrong if I can't sue. But that cannot be the answer. To the community organizer, it is this sort of legalistic approach that gives lawyers a bad name. When lawyers face such situations, they have to get creative and take their lead from the communities they serve to identify innovative solutions that are grounded in community organizing and mobilization. Reforming social media practices should face similar scrutiny, and we should not back down because there is not a ready solution, grounded in technology alone, for the problems those practices generate.

What slacktivism, fake news, and the trust-in-the-platform problem reveal is that while digital tactics can both create and activate a new, synthetic social capital to help groups form, organize, and mobilize to address the collective action problems that are often the targets of social movements, such digital tools must be linked to the trust-building effects that come about by analog, "old school" organizing efforts. As the sociologist Zeynep Tufekci argues, digitally enhanced networks must build on their early successes by creating effective, trust-generating relationships that can form the basis of sustained and sustainable action.[53] Indeed, many see the success of the Obama campaign for the presidency in 2008 as a triumph of digital tools.[54] At the same time, there is much regret in progressive circles that after that victory, little effort was made to develop an organizing infrastructure that could be mobilized for sustained efforts to support the Obama agenda.[55] Digital tools need a human touch and need to be supplemented by face-to-face connections that build trust, reduce social distance, and create robust social capital networks.

The Future of Change

Harnessing the power of social media and other tools in the current social innovation moment will not be easy. And the value that those tools have in helping like find like can serve constructive and destructive ends. Just as I do not believe progressive social change follows automatically from the introduction of a new communications technology, I do not believe such change will happen if we all just organize into translocal groups and start talking to each other. There are those who need no convincing to support progressive change, those who can be convinced, and those who will not support it. Alan Jenkins describes the challenge as figuring out "how to mobilize the base, persuade the persuadables, and marginalize the opposition."[56]

In their celebrated work *Switch: How to Change Things When Change Is Hard*, the brothers Dan and Chip Heath argue that when we think about change, we should use as a metaphor the concept of an individual riding an elephant. The rider may want to go in a certain direction, but the elephant can sometimes lose focus or energy. The rider can also get confused by a lack of clear direction. They argue that change requires three things. First, that we direct the rider by giving him or her clear direction. Second, we motivate the elephant by spurring the emotions that will keep it engaged. And finally, we shape the path: make change easy by offering step-by-step guidance, translating small, isolated victories into a sustained and sustainable campaign.[57]

Perhaps it is possible that social change brought about by social movements in social innovation moments has a few, straightforward (though I will not say easy) components. The Heath brothers suggest that in order to make change one has to have such simple guidance, so I will try to offer it here. Individuals and organizations that have had some success in promoting social change in social innovation moments have utilized medium, network, and message effectively. They have done so by using the means of communication available to them to shape the contours of their efforts and garner support. They have created translocal networks that leverage social capital as it exists in such networks, forged alliances, lowered social distance, and built and increased trust. And they have promoted an inclusive message that focused on what interests different groups share, their commonalities and not their differences.

These components—the social change matrix—have thus been central to successful social movements at times of technological change. Indeed, new technologies have been at the center of this matrix and have helped shape social movements in ways that have promoted trust and social distance. At times, however, those new technologies, standing alone, may have diminished them. The communications technologies available today are helping groups to reduce social distance in ways that can replicate the benefits of face-to-face communications that were so prevalent in the first two centuries of the American experiment. Indeed, translocal, cross-class groups were the main drivers of social change in the United States for generations. It is this combination of technology, network and message that helps bring about social change, but it is technologies, networks, and messages that stress our humanity and enable us to connect on a personal level that hold out the greatest hope for change. Roughly 150 years ago, Frederick Douglass captured the essence of these connections as follows:

> The growth of intelligence, the influence of commerce, steam, wind, and lightning are our allies. It would be easy to amplify this summary, and to swell the vast conglomeration of our material forces; but there is a deeper and truer method of measuring the power of our cause, and of comprehending its vitality. This is to be found in its accordance with the best elements of human nature. It is beyond the power of slavery to annihilate affinities recognized and established by the Almighty. The slave is bound to mankind by the powerful and inextricable net-work of human brotherhood. His voice is the voice of a man, and his cry is the cry of a man in distress, and man must cease to be man before he can become insensible to that cry.[58]

To echo Douglass, and to summarize much of what has been said in just a few words: in order to bring about positive social change, *we must harness technology to create community across difference.* Recent campaigns, described in these pages, are using new types of technologies to replicate, to a certain extent, the benefits of face-to-face communications and building connections on a personal level through messages that stress shared humanity and interest convergence, lowering social distance. They are also doing the painstaking work—often aided by technology—to enhance opportunities for face-to-face communications. They are thus finding ways to use technology that can re-create, sometimes virtually, the translocal, diverse groups that helped shape social change for generations. And it is just these tactics, perhaps, that are the future of change.

Notes

INTRODUCTION

1. This account is described in Pitkin, "How the First GI Bill Was Written (Part II)," 50; and Cleland, "Gibson's Midnight Ride Saves G.I. Bill."

2. See, e.g., Ackerman, "Constitutional Politics/Constitutional Law," 477; Ackerman, *We the People*, 266–94.

3. Jenkins and Form, "Social Movements and Social Change," 331.

4. Skocpol, *States and Social Revolutions*, 4.

5. Works that address these topics include Lanier, *Ten Arguments for Deleting Your Social Media Accounts Right Now*; McNamee, *Zucked*; and Zuboff, *Age of Surveillance Capitalism*.

6. Drucker, *Post-Capitalist Society*, 3.

7. See definitions of *matrix* in the Merriam-Webster dictionary at www.Merriam-Webster.com.

8. Shirky, *Here Comes Everybody*, 17.

1. MEDIUM

1. 2 Reg. Deb. 12 (1836) (statement of Rep. John W. Jones), 2034–36.

2. Gonzalez and Torres, *News for All the People*, 47.

3. The excerpt here is by no means the only version of the Resolves printed: there were some discrepancies between the version that was passed and the one that was ultimately printed by several newspapers. "The Resolutions as Printed by the *Maryland Gazette*, July 4, 1765," reprinted in Morgan, *Prologue to Revolution*, 49–50.

4. Starr, *Creation of the Media*, 66.

5. Ibid., 66.

6. Morgan and Morgan, *Stamp Act Crisis*, 281–83.

7. Adelman, "Constitutional Conveyance of Intelligence, Public and Private," 724.

8. Ibid., 733.

9. Brown, *Revolutionary Politics in Massachusetts*, 181–84.

10. Adelman, "Constitutional Conveyance," 739.

11. Cooke, *Reporting the War*, 13–14.

12. Tebbel, *Media in America*, 44–49.

13. Kazin, Edwards, and Rothman, *Princeton Encyclopedia of American Political History*, 219.

14. *Boston Evening Post*, May 22, 1769, quoted in Miller, *Origins of the American Revolution*, 289.

15. Jensen, *Founding of a Nation*, 127–28.

16. For a description of the American Revolution as a social movement, see generally Jameson, *American Revolution Considered as a Social Movement*.

17. Ramsay, *History of the American Revolution*, 2:406.

18. On the spread of the printing press in colonial America, see Wroth, *Colonial Printer*.

19. Adelman, "Constitutional Conveyance," 712.

20. John, *Spreading the News*, 5.

21. Ibid.

22. Ibid., 3–4.
23. Ibid., 4 (emphasis added).
24. Story, *Commentaries on the Constitution*, 3:23.
25. Tocqueville, *Democracy in America*, 489.
26. Ibid.
27. Ibid., 180-81.
28. Skocpol, *Diminished Democracy,* 31.
29. W.E. Channing, *Works of William E. Channing, D.D.*, 282.
30. Ibid., 282–83.
31. Skocpol, *Diminished Democracy,* 40–41.
32. Ibid., 37–40.
33. Ibid., 38–40.
34. Ibid., 25–26, table 2.1.
35. Sklar, "Women Who Speak for an Entire Nation."
36. Stanton, *Eighty Years or More*, 53.
37. Howe, *What God Hath Wrought,* 847.
38. Anderson, *Joyous Greetings,* 22–24.
39. Coe, "Speech of Emma Coe," quoted in *Woman's Rights Convention* (proceedings, Worcester, MA, October 15 and 16, 1851),107.
40. Skocpol, *Diminished Democracy*, 48.
41. Ibid.
42. Chambers, *Tyranny of Change,* 27.
43. Ibid.,75.
44. Ibid.,31.
45. Ibid., 86.
46. Painter, *Standing at Armageddon,* 247–48.
47. Diner, *Very Different Age,* 133.
48. Ibid., 90.
49. Chambers, *Tyranny of Change,* 18.
50. Quoted in Hyland, *George Gershwin*, xii.
51. Chambers, *Tyranny of Change,* 20.
52. Diner, *Very Different Age,* 4.
53. Chambers, *Tyranny of Change,* 7.
54. Cyphers, *National Civic Federation and the Making of a New Liberalism,* 176.
55. Cherny, *Righteous Cause*, 112.
56. DuBois, *Women's Suffrage and Women's Rights*, 197.
57. Skocpol, *Diminished Democracy,* 205.
58. Ibid.
59. Ibid., 50.
60. Putnam, *Bowling Alone*, 54.
61. Lenthall, *Radio's America*, 12.
62. Ibid., 78.
63. Brown, *Manipulating the Ether*, 61–63.
64. Roosevelt, First inaugural address, March 4, 1933.
65. Brown, *Manipulating the Ether*, 61–65.
66. Roosevelt, FDR Letter to Ralph W. Farrell.
67. Ibid.
68. Brown, *Manipulating the Ether*, 65.
69. Badger, *New Deal*, 71.
70. "Banking Normal Again," *New York Times*, March 14, 1933.
71. Ibid.
72. Putnam, *Bowling Alone*, 53–54.

73. Dunn, "Letter to the President," 68.
74. Branch, *Parting the Waters*, 762.
75. Manis, *Fire You Can't Put Out*, 321, 340.
76. Bass, *Blessed Are the Peacemakers,* 96.
77. Quoted in Eskew, *But for Birmingham,* 221.
78. Garrow, *Bearing the Cross,* 247.
79. Bass, *Blessed Are the Peacemakers,* 132.
80. Branch, *Parting the Waters,* 761.
81. Quoted in Bass, *Blessed Are the Peacemakers,* 133.
82. Evers, "I Speak as a Native Mississippian."
83. Garrow, *Bearing the Cross,* 261.
84. "Johnson Conversation with Martin Luther King on Jan. 15, 1965."
85. Bodroghkozy, *Equal Time.*
86. Carter, *Politics of Rage.*
87. Hamilton, *Bench and the Ballot,* 235.
88. Quoted in Howell Raines, *My Soul Is Rested,* 136.
89. Torres, *Black, White, and in Color,* 15–18, 36–37.
90. Ibid., 36–37.
91. Russell Baker, "Stomach-Bulge Defense," *New York Times,* April 5, 1975.
92. McLuhan, *Understanding Media,* 127.

2. NETWORK

1. Risley, *Abolition and the Press,* 39.
2. Ibid., 40.
3. Magdol, *Antislavery Rank and File,* 7–8.
4. Newman, *Transformation of American Abolitionism,* 153.
5. For a description of the efforts of the agency system, see Ibid., 152–75.
6. Merrill and Ruchames, eds., *Letters of William Lloyd Garrison,* 1:163 (emphasis in original).
7. Ibid.
8. Newman, *Transformation of American Abolitionism,* 147–48 (citations omitted).
9. Magdol, *Antislavery Rank and File,* 74.
10. Newman, *Transformation of American Abolitionism,* 138–39. On the role of black abolitionists in the abolitionist cause, see generally Quarles, *Black Abolitionists.*
11. Newman, *Transformation of American Abolitionism,* 153.
12. Ibid., 173–75.
13. Dowding, *Power,* 31.
14. Aristotle, *Politics,* 57.
15. See, e.g., Dahl, *Pluralist Democracy in the United States.*
16. Olson, *Logic of Collective Action,* 57–63.
17. Hardin, "Tragedy of the Commons."
18. Ostrom, *Governing the Commons,* 101–2.
19. Della Porta and Diani, *Social Movements,* 101.
20. Olson, *Logic of Collective Action,* 61–63.
21. Ostrom, *Governing the Commons,* 101–2.
22. Ibid.
23. See Baird, Gertner, and Randal, *Game Theory and the Law,* 312–13 (describing the prisoner's dilemma). For meta-analyses of game theory studies, see Sally, "Conversation and Cooperation in Social Dilemmas," and Sally, "Game Theory Behaves."
24. I explore these insights, and others, in greater depth in "Trust in the Shadows: Law, Behavior and Financial Re-regulation."

25. Bó, "Cooperation under the Shadow of the Future"; Evans and Thomas, "Reputation and Experimentation in Repeated Games with Two Long-Run Players"; Gibbons, "Trust in Social Structures."

26. Axelrod, *Evolution of Cooperation*, 7–14 (describing experiments).

27. Rose, "Trust in the Mirror of Betrayal," 531.

28. Axelrod, *Evolution of Cooperation,* 13; Kreps, *Game Theory and Economic Modeling,* 65–89.

29. Kahan, "Logic of Reciprocity."

30. Axelrod, *Evolution of Cooperation,* 109.

31. Ibid., 113–17; Gambetta, "Can We Trust Trust?," 227.

32. Axelrod, *Evolution of Cooperation,* 31.

33. Hoffman, McCabe, and Smith, "Social Distance and Other-Regarding Behavior in Dictator Games," 658.

34. Uslaner, *Moral Foundations of Trust*, 123–24.

35. DeBruine, "Facial Resemblance Enhances Trust," 1311.

36. Bohnet and Frey, "Sound of Silence in Prisoner's Dilemma and Dictator Games," 46; Alberto, Alesina, and La Ferrara, *Determinants of Trust.*

37. Zak and Knack, "Trust and Growth," 312–13 (finding income inequality within a country is "significantly related to" a decline in trust among the population within that country.

38. Uslaner, *Moral Foundations of Trust*, 165–81.

39. Loomis, "Communication, the Development of Trust, and Cooperative Behavior," 314–15.

40. Desforges et al., "Effects of Structured Cooperative Contact on Changing Negative Attitudes toward Stigmatized Social Groups," 531.

41. Mazar, Amir, and Ariely, "Dishonesty of Honest People."

42. Putnam, "*E Pluribus Unum*," 137.

43. On social capital generally, see Putnam, *Bowling Alone*, and Coleman, "Social Capital in the Creation of Human Capital."

44. Hanifan, "Rural School Community Center," 130.

45. Putnam, "Bowling Alone," 67; Mazzone, "Toward a Social Capital Theory of Law," 6–8.

46. Putnam, "Bowling Alone," 67.

47. Ibid.

48. de Souza Briggs, "Brown Kids in White Suburbs," 178.

49. Ibid.

50. Grootaert and Bastelaer, "Understanding and Measuring Social Capital." See also Gittell and Vidal, *Community Organizing*, 10.

51. de Souza Briggs, 178.

52. World Bank, *World Development Report 2000/2001*.

53. Putnam, *Bowling Alone*, 22.

54. Servon, "Credit and Social Capital."

55. Granovetter, "Strength of Weak Ties."

56. Putnam, *Bowling Alone*, 340.

57. Glaeser et al., "Measuring Trust," 814.

58. Putnam, "*E Pluribus Unum*," 141–51.

59. Uslaner, "Segregation, Mistrust and Minorities," 416.

60. See, e.g., Hutchinson, "Preventing Balkanization or Facilitating Racial Domination."

61. For a review of the literature regarding changes in social capital and civic engagement, see, generally, Stolle and Hooghe, "Review Article: Inaccurate, Exceptional, One-Sided or Irrelevant?"

62. See, generally, Skocpol, *Diminished Democracy*.

63. Putnam, Leonardi, and Nanetti, *Making Democracy Work*. For additional insight, see Coleman, "Social Capital," S105–8; LaPorta, Lopez-de-Silane, Shleifer, and Vishny, "Trust in Large Organizations," 333; Knack and Keefer, "Does Social Capital Have an Economic Payoff?" 1275–77; Knack, "Social Capital and the Quality of Government," 772.

64. Skocpol, *Diminished Democracy*, 221–23.

65. Kahan, "Logic of Reciprocity," 71–72. See also Rawls, *Theory of Justice* 11–19. (arguing that reciprocity and cooperation are at the heart of a just society).

66. Tocqueville, *Democracy in America*, 488.

67. McCarthy and Wolfson, "Resource Mobilization by Local Social Movement Organizations."

68. Christen and Speer, "Contextual Influences on Participation in Community Organizing," 258.

69. Whitman, *Stand Up!*, 364.

70. Slaughter, *Chessboard and the Web*, 83.

71. Crutchfield, *How Change Happens*, 27.

72. Putnam, "*E Pluribus Unum*," 137.

73. Hardin, *Trust and Trustworthiness*, 83–84.

74. Putnam, *Bowling Alone*, 67.

75. Ostrom and Ahn, "Meaning of Social Capital and Its Link to Collective Action," 22.

76. Kadushin, *Understanding Social Networks*, 14.

77. See, e.g., Christakis and Fowler, "Spread of Obesity in a Large Social Network over 32 Years," 371.

78. Kadushin, *Understanding Social Networks*, 18; Hipp and Perrin, "Simultaneous Effect of Social Distance and Physical Distance on the Formation of Neighborhood Ties."

79. Kadushin, *Understanding Social Networks*, 18–21.

80. Verbrugge, "Structure of Adult Friendship Choices," 591–94; McPherson, Smith-Lovin, and Cook, "Birds of a Feather," 418–29.

81. On Metcalfe's law, see Barabási, *Network Science*, 55.

82. Gilder, "Metcalfe's Law and Legacy."

83. This law can be depicted in a mathematical formula as follows, with n as the number of nodes: $n \times (n - 1) = n^2 - n$. See Shapiro and Varian, *Information Rules*, 184.

84. Kadushin, *Understanding Social Networks*, 26–30.

85. Ibid., 29–30.

86. McCarthy and Wolfson, "Consensus Movements, Conflict Movements, and the Cooptation of Civic and State Infrastructures," 282.

87. The filling of structural holes is sometimes referred to as "clustering." See Kadushin, *Understanding Social Networks*, 122–30.

88. West, *Scale*.

89. Ibid., 103–26.

90. Ibid., 103.

91. Ibid., 128.

92. Ostrom and Ahn, "Meaning of Social Capital and Its Link to Collective Action," 25.

93. Catt and Shuler, *Woman Suffrage and Politics*, 26.

94. Wellman, *Road to Seneca Falls*, 226; Banaszak, *Why Movements Succeed or Fail*, 6–7.

95. Banaszak, *Why Movements Succeed or Fail*, 7–9.

96. Ibid., 9.

97. Ibid., 48–49.

98. Catt and Shuler, *Woman Suffrage and Politics*, 108–20.

99. Banaszak, *Why Movements Succeed or Fail*, 47–49.

100. Ibid., 65–72.

101. Winter, *Business of Being a Club Woman*, 4.

102. Ibid., 234.
103. Ibid., 235.
104. Ibid., 247.
105. Fischer, *Paul Revere's Ride*, 19.
106. Ibid., 20.
107. Ibid., 139.
108. These sorts of linked networks appear in business settings. Brown and Duguid, *Social Life of Information*, 149–51; Sabel, "Constitutional Ordering in Historical Context."
109. Gelb, "Social Movement 'Success,'" 273.
110. Freeman, "Origins of the Women's Liberation Movement," 809.
111. Ibid.

3. MESSAGE

1. Quoted in Pitkin, "How the First GI Bill Was Written," 25.
2. Ibid.
3. Ibid.
4. Ibid., 26.
5. Ibid., 27.
6. Ibid., 51.
7. Ibid., 52.
8. Ibid., 26.
9. Skocpol, "Targeting within Universalism," 414.
10. For a discussion of public opinion polling in Detroit following racial unrest there, which found there was likely broad support for efforts to improve the economic conditions of the community regardless of the race of the beneficiary, see Aberbach and Walker, "Political Trust and Racial Ideology."
11. Johnson, "Special Message to the Congress."
12. Ibid.
13. Ibid.
14. Tocqueville, *Democracy in America*, 501.
15. McCarthy and Zald, "Resource Mobilization and Social Movements," 1216.
16. Coase, "Nature of the Firm," 390–92.
17. Tarrow, *Power in Movement*, 3.
18. Della Porta and Diani, *Social Movements*, 15.
19. Ferree, "Political Context of Rationality," 31.
20. Oliver and Marwell, "Mobilizing Technologies for Collective Action," 252.
21. Buechler, "Beyond Resource Mobilization?," 231.
22. Ganz, *Why David Sometimes Wins*, 194.
23. Ibid.
24. Goffman, *Frame Analysis*, 21–28.
25. Snow and Benford, "Master Frames and Cycles of Protest."
26. Snow, Rochford, Worden, and Benford, "Frame Alignment Processes, Micromobilization, and Movement Participation."
27. Ferree and Miller, "Mobilization and Meaning," 49.
28. "Johnson Conversation with Martin Luther King on Jan. 15, 1965."
29. Bell, "*Brown v. Board of Education* and the Interest-Convergence Dilemma," 524.
30. Ibid.
31. Ibid., 524–26.
32. Ibid., 525.
33. Quoted in Roe, ed., *Speeches and Letters of Abraham Lincoln, 1832–1865*, 194–95.
34. Lo, *Small Property versus Big Government.*

35. See, e.g., Kahneman and Tversky, "Choices, Values and Frames."

36. Buechler, "Beyond Resource Mobilization?"

37. Ennis and Schreuer, "Mobilizing Weak Support for Social Movements."

38. Shalot, *Optimism Bias.*

39. Weinstein, "Unrealistic Optimism about Future Life Events."

40. Chapman and Johnson, "Incorporating the Irrelevant," 133.

41. Westen, *Political Brain*, 100.

42. McAdam, "Recruitment to High-Risk Activism," 86–87.

43. Snow, Zurcher, and Ekland-Olson, "Social Networks and Social Movements," 798.

44. Tarrow, *Power in Movement*, 21.

45. Tversky and Kahneman, "Availability."

46. Oberschall, *Social Conflict and Social Movements*, 125; McAdam, McCarthy, and Zald, "Social Movements," 715; Klandermans and Oegema, "Potentials, Networks, Motivations, and Barriers," 520.

47. Gamson, "Social Psychology of Collective Action," 60. See also Fantasia, *Cultures of Solidarity*, 235.

48. McAdam, "Culture and Social Movements," 43.

49. Morris, *Origins of the Civil Rights Movement*, 35–38.

50. Freeman, "Origins of the Women's Liberation Movement."

51. Klandermans and Oegema, "Potentials, Networks, Motivations, and Barriers," 520 (and citations found therein).

52. Jackson et al., "Failure of an Incipient Social Movement," 38.

53. McCarthy, "Pro-Life and Pro-Choice Mobilization."

54. Staggenborg, "Consequences of Professionalization and Formalization in the Pro-Choice Movement," 593–94. For a study showing mixed approaches that included media, face-to-face efforts, and the bridging of weak ties was a more effective mobilizing approach than the use of media alone, see Briët, Klandermans, and Kroon, "How Women Become Involved in the Women's Movement of the Netherlands," 52–57.

55. Gamson, "Social Psychology of Collective Action," 71–72.

56. Melucci, "Strange Kind of Newness," 117; Buechler, "Beyond Resource Mobilization?"; Johnston, Laraña, and Gusfield, "Identities, Grievances and New Social Movements," 10.

57. Taylor and Whittier, "Collective Identity in Social Movement Communities."

58. Fireman and Gamson, "Utilitarian Logic in the Resource Mobilization Perspective," 21.

59. Snow, Rochford, Worden, and Benford, "Frame Alignment Processes, Micromobilization, and Movement Participation," 467.

60. McAdam, "Culture and Social Movements," 53.

61. Tarrow, *Power in Movement*, 5.

62. Whitman, *Stand Up!*, chapter 5.

63. Tarrow, *Power in Movement*, 7.

64. Ibid., 21.

65. Buechler, "Beyond Resource Mobilization?," 222.

66. Alinsky, *Rules for Radicals*, 20.

4. THE GREAT DIVIDE

1. Richard A. Viguerie, interview by Ray Brescia, Dec. 9, 2014.Richard A. interview by Raymond H. Brescia, Dec. 9, 2014. pp.t citation but make it an endnote.. Endnotes should be in MLA sty Unless otherwise noted, all quotes from Viguerie are taken from this interview.

2. Viguerie and Franke, *America's Right Turn*; Viguerie interview.Richard A. interview by Raymond H. Brescia, Dec. 9, 2014. pp.t citation but make it an endnote.. Endnotes should be in MLA sty

3. Huckshorn, "National Committee Leadership of State and Local Parties," 41.

4. Viguerie, interview.

5. Klinkner, "A Comparison of Out-Party Leaders, "137.

6. Polsby, *Presidential Elections*, 56.

7. Quoted in Sandoz and Crabb, Jr., eds., *A Tide of Discontent*, 14.

8. Bailey, *The Republican Party in the U.S. Senate*, 3–4.

9. Ibid., 41–42.

10. Viguerie and Franke, 108.

11. Ibid.

12. Viguerie, interview.

13. Lepore, *These Truths*, 665 (citation omitted).

14. Viguerie and Franke, 325.

15. Putnam, *Bowling Alone*, 54–55.

16. Ibid.

17. Ibid., 55.

18. Skocpol, *Diminished Democracy*, 130–31.

19. Almond and Verba, *The Civic Culture*, 306.

20. Putnam, *Bowling Alone*, 47.

21. Ibid., 17.

22. Oliver and Marwell, "Mobilizing Technologies for Collective Action," 270.

23. Handler, *Social Movements and the Legal System*, 1.

24. Powell, Jr., "Attack on American Free Enterprise System," 1.

25. Ibid.

26. Ibid., 2.

27. Ibid.

28. Ibid., 2–3.

29. Ibid.

30. Ibid., 12.

31. Ibid.

32. Ibid., 30.

33. Ibid.

34. Ibid.

35. Ibid.

36. Judis, *The Paradox of American Democracy,*125–27.

37. Skopcol, 154–55.

38. "Union Members Summary," U.S. Dept. of Labor Bureau of Labor Statistics.

39. Minkoff, *Organizing for Equality*, 61–64.

40. Lepore, 662.

41. Mansbridge, *Why We Lost the ERA*, 172.

42. Ibid.

43. Ibid., 172–73.

44. Ibid., 171.

45. Gelb and Palley, *Women and Public Policies*, 25.

46. Mansbridge, 174–75.

47. Ibid., 159.

48. Quoted in Mansbridge, 159.

49. Ibid.

50. Ibid.

51. Felsenthal, *The Sweetheart of the Moral Majority*, 267–68.
52. Mansbridge, 159.
53. Ibid., 44.
54. Gelb and Palley, *Women and Public Policies*, 5.
55. Ibid., 6.
56. McCarthy and Zald, *The Trend of Social Movements in America*, 20.
57. Oliver and Marwell, "Mobilizing Technologies for Collective Action," 260.
58. Handler, 8.
59. Weir and Ganz, "Reconnecting People and Politics," 160.
60. Skocpol, 174.
61. On the relationship between trust and inequality, see Uslaner, *The Moral Foundations of Trust*, 186–188.
62. On inequality in the United States from the Great Depression to contemporary times, see Lepore, *These Truths*, 757. For an exploration of some of the sources of the decline in trust, see Campbell, "Tax Revolts and Political Change," 164–165. For an analysis of the causes of economic inequality in the United States, Timothy Noah, *The Great Divergence*, 60–143.
63. Tocqueville, *Democracy in America*, 3.
64. Ibid., 490.
65. Piketty, Saez, and Zucman. "Distributional National Accounts."

5. DIGITAL ORGANIZING

1. Emily Comer and Jay O'Neal, interview by Ray Brescia, April 10, 2018. All quotes from Comer and O'Neal are taken from this interview.
2. Ibid.
3. Ibid.
4. See, e.g., Putnam, *Bowling Alone*, 135–41.
5. Ibid., 140–47. On the interplay between trust and social capital, see Norris, *Democratic Phoenix*, 153–56.
6. Putnam, *Bowling Alone*, 283–84.
7. Niraj Chokshi, "Out of Office: More People Working Remotely, Survey Finds," *New York Times*, February 15, 2017.
8. Springsteen, "57 Channels (and Nothin' On)," *Human Touch*, 1992.
9. McLuhan, *Understanding Media*, 37–50.
10. See, e.g., Wyrwoll, "Social Media," 12–24; and Shirky, *Cognitive Surplus*, 51–55.
11. Standage, *Writing on the Wall*, 213.
12. Palfrey and Gasser, *Born Digital*, 1–7.
13. Barnitt, Chadwick, and Whitney, "Creating a Social Media Movement," 41–48.
14. Cowen, *Complacent Class*, 13–15.
15. Sunstein, *#Republic*, 59–61.
16. Perry and Ricca, "Instant Messaging," 211–12.
17. Benkler, *Wealth of Networks*, 106–29.
18. See, generally, Starr, *Creation of the Media*.
19. Benkler, *Wealth of Networks*, 30.
20. Friedman, *Thank You for Being Late*, 19–21 (describing introduction of the iPhone and other events that occurred in 2007).
21. Bounanno, Montolio, and Vanin, "Does Social Capital Reduce Crime?," 150.
22. Halbert, "Two Faces of Disintermediation," 86–89.
23. Quan-Haase and Wellman, "How Does the Internet Affect Social Capital?," 126.
24. Hampton and Wellman, "Neighboring in Netville," 303.
25. Ibid., 303.

26. Obar, Zube, and Lampe, "Advocacy 2.0," 12–18.
27. Tufekci, *Twitter and Tear Gas*, 269–70.
28. Ibid., 3–27.
29. Henkin, *Postal Age*, 18.
30. Standage, *Victorian Internet*, 63.
31. Margetts and John, *Political Turbulence*, 196–97.
32. Benkler, *Wealth of Networks*, 30.
33. Perry Stein, "The Woman Who Started the Women's March with a Facebook Post Reflects: 'It Was Mind-Boggling.'" *Washington Post*, January 31, 2017.
34. Megan Farokhmanesh, "The Positive Peer Pressure of the "I Voted" Selfie." *The Verge*, November 8, 2016, accessed April 4, 2019, https://www.theverge.com/2016/11/8/13552446/election-day-i-voted-sticker-selfie-social-media.
35. Bennett and Segerberg, *Logic of Connective Action*, 53.
36. Oscar Wilde famously quipped: "The trouble with Socialism is that it takes too many evenings"; quoted in Purdy, *For Common Things*, 68.
37. Margetts and John, *Political Turbulence*, 12.
38. Chwe, *Rational Ritual*, 3.
39. Tilly, "Repertoires of Contention in America and Britain, 1750–1830," 131.
40. Tilly, "Web of Contention in Eighteen-Century Cities," 40.
41. Davis and Zald, "Social Change, Social Theory, and the Convergence of Movements and Organizations," 347.
42. Sajuria et al., "Tweeting Alone?"
43. Sabatini and Sarracino, "E-participation, Social Capital and the Internet," 36.
44. Wellman, "Computer Networks as Social Networks," 2032.
45. Ellison, Steinfield, and Lampe, "Benefits of Facebook 'Friends,'" 1153–65.
46. Hampton and Wellman, "Neighboring in Netville," 294.
47. Kavanaugh et al., "Weak Ties in Networked Communities," 283.
48. Bimber, Flanagin, and Stohl, *Collective Action in Organizations*, 139.
49. Quan-Haase and Wellman, "How Does the Internet Affect Social Capital?," 130.
50. Tocqueville, *Democracy in America*, 498.
51. Ostrom and Ahn, "Meaning of Social Capital and Its Link to Collective Action," 28.
52. Kittilson and Dalton, "Virtual Civil Society" 641.
53. Ibid.
54. Bond et al., "61-Million Person Experiment in Social Influence and Political Mobilization," 295–98.
55. Jones et al., "Social Influence and Political Mobilization."
56. See, e.g., Desforges et al., "Effects of Structured Cooperative Contact on Changing Negative Attitudes toward Stigmatized Social Groups"; Orbell, Van de Kragt, and Dawes, "Explaining Discussion-Induced Cooperation."
57. Bennett and Segerberg, *Logic of Connective Action*, 33.
58. McLuhan, *Understanding Media*, 5–7.
59. Lundby, "Introduction: Digital Storytelling, Mediated Stories," 4.
60. Ostrom, *Governing the Commons*, 83.
61. Wood, *Creation of the American Republic* 1776–1787, 24.
62. Taylor, *People's Platform*, 2.
63. Surowiecki, *Wisdom of Crowds*, 29–30.
64. Dorf and Sabel, "Constitution of Democratic Experimentalism," 288–89.
65. Surowiecki, *Wisdom of Crowds*, xiv. See also Howe, *Crowdsourcing*, 14.
66. Tocqueville, *Democracy in America*, 497
67. Ibid., 181.
68. Whitman, *Stand Up!*, Introduction.

69. Shirky, *Here Comes Everybody*, 47.

70. Ostrom, *Governing the Commons*, 83.

71. Tyler, *Why People Obey the Law*, 163.

72. Taleb, *Skin in the Game.*

73. Ostrom, *Governing the Commons*, 83. See also Sabel, "Constitutional Ordering in Historical Context," 92–94.

74. See, e.g., Liebman and Sabel, "Public Laboratory Dewey Barely Imagined," 301–4.

75. Charles Krauthammer, "Be Afraid: The Meaning of Deep Blue's Victory," *Weekly Standard*, May 26, 1997; Kasparov, *Deep Thinking*, 167–96.

76. Kelly, *Inevitable*, 41.

77. Brynjolfsson and McAfee, *Second Machine Age*, 284.

6. AMENDING THE VIOLENCE AGAINST WOMEN ACT

1. For background on the VAWA reauthorization, see, e.g., Suzy Khimm, "The Violence Against Women Act Is on Life Support," *Washington Post*, January 25, 2013.

2. For some of the progressive critiques of VAWA, see, e.g., Kate Pickert, "What's Wrong with the Violence Against Women Act?," *The Nation*, February 27, 2013. For a description of some of the conservative critiques of VAWA, see Molly Ball, "Why Would Anyone Oppose the Violence Against Women Act?," *The Atlantic*, February 12, 2013.

3. Sharon Stapel, interview by Ray Brescia, April 2, 2014. Unless otherwise specified, all quotes from Stapel are from this interview.

4. For more information on the exclusion of LGBTQ individuals from VAWA and other concerns with the violence "against women" terminology and framework, see Goldscheid, "Gender Neutrality, the Violence against Women Frame, and Transformative Reform."

5. For more on intimate partner violence, the LGBTQ community, and VAWA, see National Coalition of Anti-Violence Programs, *Lesbian, Gay, Bisexual, Transgender, Queer and HIV-Affected Intimate Partner Violence.*

6. For some of the problems related to prosecution of intimate partner violence crimes perpetrated against Native American women, see Mullen, "Violence Against Women Act." On intimate partner violence, sexual assault, the Native American community, and the barriers to obtaining justice, see Amnesty International, *Maze of Injustice.*

7. Rosie Hidalgo, interview by Ray Brescia, February 12, 2014. All subsequent quotes from Hidalgo are from this interview.

8. Pat Reuss, interview by Ray Brescia, February 3, 2014. All subsequent quotes from Reuss are from this interview.

9. For more on U visas and the ways in which intimate partner violence affects the immigrant community, see Olivares, "Battered by Law." For a description of violence against Latina immigrants, see Villalón, *Violence against Latina Immigrants.* See also Villalón, "Violence against Immigrants in a Context of Crisis."

10. For a description of an earlier version of a reauthorization bill that passed the House, see Jennifer Bendery and Laura Bassett, "House Passes Violence Against Women Act That Leaves Out LGBT, Immigrant Protections," *Huffington Post*, May 16, 2012, accessed April 1, 2019, https://www.huffingtonpost.com/2012/05/16/house-passes-violence-against-women-act_n_1522524.html.

11. John Eligon and Michael Schwirtz, "Senate Candidate Provokes Ire with 'Legitimate Rape' Comment," *New York Times*, August 19, 2012.

12. For a discussion of some of the reasons Republican candidates faced challenges attracting women voters in the 2012 election cycle, see Frank Rich, "Stag Party: The GOP's Woman Problem Is That It Has a Serious Problem with Women," *New York Magazine*, March 25, 2012.

13. Sharon Stapel, "The Violence Against Women Act and Why Language Matters," *TakePart.com,* December 19, 2012, accessed April 1, 2019, https://news.yahoo.com/op-ed-violence-against-women-act-why-language-180058869.html.

14. For an overview of the 2013 changes to VAWA, see National Task Force to End Sexual and Domestic Violence Against Women, *Summary of Changes from VAWA Reauthorization 2013.*

7. MARRIAGE EQUALITY IN MAINE

1. Wolfson, "Samesex Marriage and Morality," 2.

2. Ibid., 3.

3. Ibid., 31.

4. Evan Wolfson, interview by Ray Brescia, February 5, 2014. Unless otherwise specified, all quotations from Wolfson are from this interview.

5. See Waddock, Waddell, and Gray, "Transformational Change Challenge of Memes."

6. Molly Ball, "The Marriage Plot: Inside This Year's Epic Campaign for Gay Equality," *The Atlantic,* December 11, 2012.

7. Andrew Sullivan, "Here Comes the Groom: A (Conservative) Case for Gay Marriage," *New Republic,* August 28, 1989.

8. Sullivan, *Virtually Normal,* 112.

9. Ibid., 133–36.

10. Amy Mello, interview by Ray Brescia, February 11, 2014. All quotations from Mello are from this interview.

11. For a transcript of this interview, see "Transcript: President Obama's ABC News Interview on Same-Sex Marriage," *Sojourners,* May 10, 2012, accessed April 4, 2019 https://sojo.net/articles/transcript-president-obamas-abc-news-interview-same-sex-marriage.

12. Braunach, "Changing Same-Sex Marriage Attitudes in America from 1988 through 2010."

13. Balkin, "*Brown,* Social Movements and Social Change," 246–47.

14. Debra Cassens Weiss, "Justice Ginsburg: Roe v. Wade Decision Came Too Soon," *ABA Journal,* February 13, 2012.

15. That is not to say that there has not been resistance to the ultimate outcome of the marriage equality campaign, as local government officials have sometimes refused to certify same-sex marriages, or, more recently, when a private commercial establishment was given license by the Supreme Court to discriminate against a gay couple. See Alan Blinder and Richard Fausset, "Kentucky Clerk Who Said 'No' to Gay Couples Won't Be Alone in Court," *New York Times,* September 2, 2015; Adam Liptak, "In Narrow Decision, Supreme Court Sides with Baker Who Turned Away Gay Couple," *New York Times,* June 4, 2018. Unless there is a dramatic shift in support for marriage equality at the Supreme Court, which is certainly not out of the question, marriage equality would seem fixed in the nation's constitutional firmament, at least for now.

16. For a description of the broad legal team responsible for litigating Obergefell before the U.S. Supreme Court, see Frank, *Awakening* 334–41.

17. Obergefell v. Hodges, 576 U.S. ___ (2015); Slip Op. at 28.

18. Ibid., at 10.

19. Wolfson, "Samesex Marriage and Morality," 77.

8. A LIVING WAGE IN LONG BEACH

1. Cited in Los Angeles Alliance for a New Economy, "Tale of Two Cities," 4.

2. Ibid.

3. Berube and Katz, *Katrina's Window,* 3.

4. Los Angeles Alliance for a New Economy, "Tale of Two Cities," 21.

5. Ibid.

6. Ibid.

7. Leigh Shelton, interview by Ray Brescia, May 8, 2014. All quotations from Shelton are taken from this interview.

8. Jeanine Pearce, interview by Ray Brescia, May 29, 2014. All quotations from Pearce are taken from this interview.

9. Loraina Lopez Masoumi, interview by Ray Brescia, May 8, 2014. All quotations from Lopez Masoumi are taken from this interview.

10. Los Angeles Alliance for a New Economy, "Raise LA."

11. Minimum Wage Study Commission, *Report of the Minimum Wage Study Commission.*

12. Brown, Gilroy, and Kohen, "Effect of the Minimum Wage on Employment and Unemployment." The authors noted that research of the impact of increases to the minimum wage on teenagers was, at the time, more extensive than research conducted on its impact on other age groups. Ibid., 47.

13. Card and Krueger, "Minimum Wage and Employment."

14. Schmitt, *Why Does the Minimum Wage Have No Discernible Effect on Employment?*

15. Autor, Manning, and Smith, *Contribution of the Minimum Wage to U.S. Wage Inequality over Three Decades.*

16. Elwell, *Inflation and the Real Minimum Wage.*

17. DiNardo, Fortin, and Limieux, "Labor Market Institutions and the Distribution of Wages, 1973–1992."

18. Dube, "Minimum Wages and the Distribution of Family Incomes."

19. Hall and Cooper, "How Raising the Federal Minimum Wage Would Help Working Families and Give the Economy a Boost."

20. Allegretto et al., *New Wave of Local Minimum Wage Policies*; Cengiz et al., "Effect of Minimum Wages on Low-Wage Jobs."

21. Piketty, *Capital in the 21st Century*, 676–91.

22. Lowrey, *Give People Money.*

9. PUTTING THE MATRIX TO WORK

1. Mettler, *Soldiers to Citizens*, 7.

2. Ibid.

3. Castells, *Rise of the Network Society,* 1:5.

4. Topolsky, "Common Cause?," 37.

5. Alan Jenkins, interview by Ray Brescia, April 30, 2013. All quotations from Jenkins are taken from this interview.

6. Brynjolfsson and McAfee, *Second Machine Age*, 188.

7. Bond et al., "61-Million Person Experiment in Social Influence and Political Mobilization."

8. Skocpol, *Diminished Democracy*, 233–35 (discussing focus of professionalized groups on cultivating donors through targeted, narrow direct mail campaigns).

9. Ibid., 108–13 (describing cross-class characteristics of many national federations in the United States prior to the 1970s).

10. Crenshaw, "Mapping the Margins."

11. Sunstein, *How Change Happens*, 35.

12. Lilla, *Once and Future Liberal*, 7.

13. Ibid., 9.

14. Ostrom and Ahn, "Meaning of Social Capital and Its Link to Collective Action."

15. Castells, *Rise of the Network Society*, 469.

16. Peter Wallsten, "NAACP Endorses Same-Sex Marriage," *Washington Post*, May 19, 2012.

17. Bond, "Address to Freedom Summer 50th Commemoration," 1.

18. Arsenault, *Freedom Riders*, 2–3.

19. Tocqueville, *Democracy in America*, 3.

20. Ibid., 490.

21. Schwarz and Paul, "Resource Mobilization versus the Mobilization of People," 206.

22. Alinsky, *Rules for Radicals*, xix.

23. Ibid., 184.

24. Ibid., 77.

25. Ibid., 77-78.

26. McAlevey, *No Shortcuts*, 23.

27. Ian Haney López, Anat Shenker-Osario, and Tamara Draut, "Democrats Can Win by Tackling Race and Class Together. Here's Proof," *The Guardian*, April 14, 2018.

28. Binyamin Appelbaum, "Face to Face with the Fed, Workers Ask for More Help," *New York Times*, November 14, 2014.

29. Ady Barkan, interview by Ray Brescia, February 22, 2018.

30. "Man with ALS Confronts Senator: 'You Can Save My Life,'" CNN, YouTube, posted December 12, 2017, https://www.youtube.com/watch?v=jUgKVoIv8Ss.

31. Barkan interview.

32. Ibid.

33. Ibid.

34. Lilla, *Once and Future Liberal*, 87.

35. See Clyde Haberman, "Philanthropy That Comes from a Click," *New York Times*, November 13, 2016.

36. Morozov, *Net Delusion*, 189–91.

37. Malcolm Gladwell, "Small Change: Why the Revolution Will Not Be Tweeted," *The New Yorker*, October 4, 2010.

38. Bill Wasik, "#Riot: Self-Organized, Hyper-Networked Revolts—Coming to a City near You," *Wired*, December 16, 2011.

39. Frenkel and Benner, "To Stir Discord in 2016, Russians Turned Most Often to Facebook."

40. Lazer et al., "Science of Fake News," 1094–96.

41. Foer, *World without Mind*, 7.

42. Mannheim, *Man and Society*, 60. Endnote omitted.

43. See, generally, Mill, *On Liberty*.

44. O'Neil, *Weapons of Math Destruction*, 185.

45. Centola, Willer, and Macy, "Emperor's Dilemma."

46. Benkler, Faris, and Roberts, *Network Propaganda*, 386.

47. For a description of Cambridge Analytica's activities in the lead-up to the 2016 election, see Information Commissioner's Office, *Investigation into the Use of Data Analytics in Political Campaigns*.

48. Sheera Frenkel and Katie Benner, "To Stir Discord in 2016, Russians Turned Most Often to Facebook," *New York Times*, February 17, 2018.

49. "Mark Zuckerberg, in His Own Words: The CNN Interview," March 21, 2018, CNN, accessed April 1, 2019, https://money.cnn.com/2018/03/21/technology/mark-zuckerberg-cnn-interview-transcript/index.html.

50. O'Shaughnessy, *Marketing the Third Reich*, 198–99.

51. McAfee and Brynjolfsson, *Machine, Platform, Crowd*, 163.

52. Ibid.

53. See Tufekci, *Twitter and Tear Gas*, 268–72.
54. Issenberg, *Victory Lab*, 246–71.
55. Micah L. Sifry, "Obama's Lost Army," *The New Republic*, February 9, 2017.
56. Jenkins interview.
57. Heath and Heath, *Switch*, 240–45.
58. Douglass, *Narrative of the Life of Frederick Douglass and Other Works*, 7644.

Bibliography

Aberbach, Joel D., and Jack L. Walker. "Political Trust and Racial Ideology." *American Political Science Review* 64, no. 4 (1970): 1199–1219.

Ackerman, Bruce. "Constitutional Politics / Constitutional Law." *Yale Law Journal* 99, no. 3 (1989): 453–547.

Ackerman, Bruce. *We the People: Foundations.* Cambridge, MA: Harvard University Press, 1991.

Adelman, Joseph M. "A Constitutional Conveyance of Intelligence, Public and Private: The Post Office, the Business of Printing, and the American Revolution." *Enterprise & Society* 11, no. 4 (2010): 709–52.

Alberto, Alesina, and Elaina La Ferrara. "Determinants of Trust." Working paper 7621, National Bureau of Economic Research, 2000.

Alinsky, Saul. *Rules for Radicals: A Pragmatic Primer for Realistic Radicals.* New York: Vintage Books, 1971.

Allegretto, Sylvia, Anna Godoey, Carl Nadler, and Michael Reich. *The New Wave of Local Minimum Wage Policies: Evidence from Six Cities.* Center on Wage and Employment Dynamics, University of California Berkeley, September 6, 2018.

Almond, Gabriel A., and Sidney Verba. *The Civic Culture: Political Attitudes and Democracy in Five Nations.* Princeton, NJ: Princeton University Press, 1963.

Altschuler, Glenn, and Stuart Blumin. *The GI Bill: The New Deal for Veterans.* Oxford: Oxford University Press, 2009.

Amnesty International. *Maze of Injustice: The Failure to Protect Indigenous Women from Sexual Violence in the USA.* Accessed March 30, 2019. https://www.amnestyusa.org/wp-content/uploads/2017/05/mazeofinjustice.pdf.

Anderson, Bonnie S. *Joyous Greetings: The First International Women's Movement, 1830–1860.* Oxford: Oxford University Press, 2001.

Aristotle. *Politics.* Translated by Benjamin Jowett. Mineola, NY: Dover, 2000.

Arsenault, Raymond. *Freedom Riders: 1961 and the Struggle for Racial Justice.* New York: Oxford University Press, 2016.

Autor, David, Alan Manning, and Christopher L. Smith. *The Contribution of the Minimum Wage to U.S. Wage Inequality over Three Decades: A Reassessment.* Finance and Economics Discussion Series, Divisions of Research & Statistics and Monetary Affairs. Washington, DC: Federal Reserve Board, 2010.

Axelrod, Robert. *The Evolution of Cooperation.* Cambridge, MA: Basic Books, 1984.

Badger, Anthony J. *The New Deal: The Depression Year, 1933–1940.* Chicago: Ivan R. Dee, 1989.

Bailey, Christopher J. *The Republican Party in the U.S. Senate: 1974–1984: Party Change and Institutional Development.* Manchester: Manchester University Press, 1988.

Baird, Douglas G., Robert Gertner, and Picker Randal. *Game Theory and the Law.* Cambridge, MA: Harvard University Press, 1998.

Balkin, Jack M., "*Brown,* Social Movements, and Social Change." In *Choosing Equality: Essays and Narratives on the Desegregation Experience*, edited by Robert L. Hayman, Jr., and Leland Ware, 246–55. University Park: Pennsylvania State University Press, 2009.

Banaszak, Lee Ann. *Why Movements Succeed or Fail: Opportunity, Culture, and the Struggle for Woman Suffrage*. Princeton, NJ: Princeton University Press, 1996.

Barabási, Albert-László. *Network Science*. Cambridge: Cambridge University Press, 2016.

Barnitt, John, Sarah Chadwick, and Sofie Whitney. "Creating a Social Media Movement: Mid to Late February." In *Glimmer of Hope: How Tragedy Sparked a Movement*, by Adam Alhanti, Dylan Baierlein, John Barnitt, Alfonso Calderon, Sarah Chadwick, Jaclyn Corin, Matt Deitsch, et al., 39–69. United States: Razorbill and Dutton, 2018.

Bass, S. Johnathan. *Blessed Are the Peacemakers: Martin Luther King, Jr., Eight White Religious Leaders, and the "Letter from Birmingham Jail."* Baton Rouge: Louisiana State University Press, 2001.

Bell, Derrick A. "*Brown v. Board of Education* and the Interest-Convergence Dilemma." *Harvard Law Review* 93, no. 3 (1980): 518–34.

Benkler, Yochai. *The Wealth of Networks: How Social Production Transforms Markets and Freedom*. New Haven, CT: Yale University Press, 2006.

Benkler, Yochai, Robert Faris, and Hal Roberts. *Network Propaganda: Manipulation, Disinformation, and Radicalization in American Politics*. New York: Oxford University Press, 2018.

Bennett, W. Lance, and Alexandra Segerberg. *The Logic of Connective Action*. Cambridge, MA: Harvard University Press, 2012.

Berube, Alan, and Bruce Katz. *Katrina's Window: Confronting Concentrated Poverty across America (2005)*. Accessed March 30, 2019. https://www.brookings.edu/wp-content/uploads/2016/06/20051012_Concentratedpoverty.pdf.

Bimber, Bruce, Andrew J. Flanagin, and Cynthia Stohl. *Collective Action in Organizations: Interaction and Engagement in an Era of Technological Change*. New York: Cambridge University Press, 2012.

Bó, Pedro Dal. "Cooperation under the Shadow of the Future: Experimental Evidence from Infinitely Repeated Games." *American Economic Review* 95, no. 5 (2005): 1591–1604.

Bodroghkozy, Aniko. *Equal Time: Television and the Civil Rights Movement*. Chicago: University of Illinois, 2012.

Bohnet, Iris, and Bruno S. Frey. "The Sound of Silence in Prisoner's Dilemma and Dictator Games." *Journal of Economic Behavior and Organization* 38 (1999): 43–57.

Bond, Julian. "Address to Freedom Summer 50th Commemoration." June 28, 2014, Jackson, MS. Accessed April 3, 2019. http://www.crmvet.org/comm/140628_fs50_bond.pdf.

Bond, Robert M., Christopher J. Fariss, Jason J. Jones, Adam D.I. Kramer, Cameron Marlow, Jaime E. Settle, and James Fowler. "A 61-Million Person Experiment in Social Influence and Political Mobilization." *Nature* 489 (2012): 295–98.

Bounanno, Paolo, Daniel Montolio, and Paolo Vanin. "Does Social Capital Reduce Crime?" *Journal of Law and Economics* 52, no. 1 (2009): 145–70.

Branch, Taylor. *Parting the Waters: America in the King Years 1954–1963*. New York: Simon & Schuster, 1988.

Braunach, Dawn Michelle. "Changing Same-Sex Marriage Attitudes in America from 1988 through 2010." *Public Opinion Quarterly* 76, no. 2 (2012): 364–78.

Brescia, Raymond H. "Trust in the Shadows: Law, Behavior and Financial Re-regulation." *Buffalo Law Review* 57, no. 5 (2009): 1361–1445.

Briët, Martien, Bert Klandermans, and Frederike Kroon. "How Women Become Involved in the Women's Movement of the Netherlands." In *The Women's*

Movements of the United States and Western Europe: Consciousness, Political Opportunity, and Public Policy, edited by Mary Feinsod Katzenstein and Carol McClurg Mueller, 44–63. Philadelphia: Temple University Press, 1992.

Brown, Charles, Curtis Gilroy, and Andrew Kohen. "The Effect of the Minimum Wage on Employment and Unemployment." *Journal of Economic Literature* 20, no. 2 (1982): 487–528.

Brown, John Seely, and Paul Duguid. *The Social Life of Information*. Boston: Harvard Business School Press, 2000.

Brown, Richard D. *Revolutionary Politics in Massachusetts: The Boston Committee of Correspondence and the Towns 1772–1774*. New York: Norton, 1970.

Brown, Robert J., *Manipulating the Ether: The Power of Broadcast Radio in Thirties America*. Jefferson, NC: McFarland, 1998.

Brynjolfsson, Erik, and Andrew McAfee. *The Second Machine Age: Work, Progress, and Prosperity in a Time of Brilliant Technologies*. New York: W.W. Norton, 2014.

Buechler, Steven M. "Beyond Resource Mobilization? Emerging Trends in Social Movement Theory." *Sociological Quarterly* 34, no. 2 (1993): 217–35.

Campbell, Ballard C. "Tax Revolts and Political Change." In *Loss of Confidence: Politics and Policy in the 1970s*, edited by David Brian Robertson, 153–78. University Park: Penn State University Press, 1998.

Card, David, and Alan Krueger. "Minimum Wage and Employment: A Case Study of the Fast-Food Industry in New Jersey and Pennsylvania." *American Economic Review* 84, no. 4 (1994): 772–93.

Carter, Dan T. *The Politics of Rage: George Wallace, the Origins of the New Conservatism, and the Transformation of American Politics*. 2nd ed. Baton Rouge: Louisiana State University Press, 2000.

Castells, Manuel. *The Rise of the Network Society*, Volume 1. Malden, MA: Blackwell, 1996.

Catt, Carrie Chapman, and Nettie Rogers Shuler. *Woman Suffrage and Politics*. Seattle: University of Washington Press, 1969.

Cengiz, Doruk, Arindrajit Dube, Attila Lindner, and Ben Zipperer. "The Effect of Minimum Wages on Low-Wage Jobs: Evidence from the United States Using a Bunch Estimator." CEP discussion paper 1531, Centre for Economic Performance, February 2018. Accessed March 30, 2019, http://cep.lse.ac.uk/pubs/download/dp1531.pdf.

Centola, Damon, Robb Willer, and Michael Macy. "The Emperor's Dilemma: A Computational Model of Self-Enforcing Norms." *American Journal of Sociology* 110, no. 4 (2005): 1009–40.

Chambers, John Whiteclay, II. *The Tyranny of Change: America in the Progressive Era 1890–1920*. 2nd ed. New Brunswick, NJ: Rutgers University Press, 2000.

Channing, W.E. *The Works of William E. Channing, D.D.* 6th ed. London: Routledge, 1846.

Chapman, Gretchen B., and Eric J. Johnson. "Incorporating the Irrelevant: Anchors in Judgments of Belief and Value." In *Heuristics and Biases: The Psychology of Intuitive Judgment*, edited by Thomas Gilovich, Dale Griffin, and Daniel Kahneman, 120-38. Cambridge: Cambridge University Press, 2002.

Cherny, Robert W. *The Righteous Cause: The Life of William Jennings Bryan*. Boston: Little, Brown, 1994.

Christakis, Nicholas A., and James H. Fowler. "The Spread of Obesity in a Large Social Network over 32 Years." *New England Journal of Medicine* 357, no. 4 (2007): 370–79.

Christen, Brian D., and Paul W. Speer. "Contextual Influences on Participation in Community Organizing: A Multi-Level Longitudinal Study." *American Journal of Community Psychology* 47, no. 3–4 (2011): 253–63.

Chwe, Michael Suk-Young. *Rational Ritual: Culture, Coordination, and Common Knowledge*. Princeton, NJ: Princeton University Press, 2013.

Cleland, Max. "Gibson's Midnight Ride Saves G.I. Bill." *Rockmart Journal*, May 6, 1992.

Coase, R.H. "The Nature of the Firm." *Economica* 84, no. 4 (1937): 386–405.

Coe, Emma R. "Speech of Emma Coe to the 1851 Women's Rights Convention, Assembled in Worcester, MA." In *The Proceedings of the Woman's Rights Convention, Held at Worcester, October 15th and 16th, 1851*, 104–9. New York: Fowlers and Wells, 1852.

Coleman, James S. "Social Capital in the Creation of Human Capital." *American Journal of Sociology* 94 (1988): S95–S120.

Cooke, John Byrne. *Reporting the War: Freedom of the Press from the American Revolution to the War on Terrorism*. New York: St. Martin's Griffin, 2008.

Cowen, Tyler. *The Complacent Class: The Self-Defeating Quest for the American Dream*. New York: St. Martin's Press, 2017.

Crenshaw, Kimberlé. "Mapping the Margins: Intersectionality, Identity Politics, and Violence against Women of Color." *Stanford Law Review* 43, no. 6 (1991): 1241–99.

Crutchfield, Leslie R. *How Change Happens: Why Some Social Movements Succeed and Others Do Not*. Hoboken, NJ: John Wiley & Sons, 2018.

Cyphers, Christopher J. *The National Civic Federation and the Making of a New Liberalism, 1900–1915*. Westport: Praeger, 2002.

Dahl, Robert A. *Pluralist Democracy in the United States*. Chicago: Rand McNally, 1967.

Davis, Gerald F., and Mayer N. Zald. "Social Change, Social Theory, and the Convergence of Movements and Organizations." In *Social Movements in Organization Theory*, edited by Gerald F. Davis, Doug McAdam, W. Richard Scott, and Mayer N. Zald, 335–50. New York: Cambridge University Press, 2005.

de Souza Briggs, Xavier. "Brown Kids in White Suburbs: Housing Mobility and the Many Faces of Social Capital." *Housing Policy Debate* 9, no. 1 (1998): 177–221.

DeBruine, Lisa M. "Facial Resemblance Enhances Trust." *Proceedings of the Royal Society of London Series B: Biological Sciences* 269, no. 1498 (2011): 1307–12. https://doi.org/10.1098/rspb.2002.2034.

Della Porta, Donatella, and Mario Diani. *Social Movements: An Introduction*. Malden, MA: Blackwell, 2006.

Desforges, Donna M., Charles G. Lord, Shawna L. Ramsey, Julie A. Mason, Marilyn D. Van Leeuwen, Sylvia C. West, and Mark R. Lepper. "Effects of Structured Cooperative Contact on Changing Negative Attitudes toward Stigmatized Social Groups." *Journal of Personality and Social Psychology* 60, no. 4 (1991): 531–44. http://dx.doi.org/10.1037/0022-3514.60.4.531.

DiNardo, John, Nicole M. Fortin, and Thomas Limieux. "Labor Market Institutions and the Distribution of Wages, 1973–1992: A Semi-Parametric Approach." *Econometrica* 64, no. 5 (1996): 1001–44.

Diner, Steven J. *A Very Different Age: Americans in the Progressive Era*. New York: Hill and Wang, 1998.

Dorf, Michael C., and Charles F. Sabel. "A Constitution of Democratic Experimentalism." *Columbia Law Review* 98, no. 2 (1998) 267–473.

Douglass, Frederick. *Narrative of the Life of Frederick Douglass and Other Works*. San Diego: Canterbury Classics, 2014.

Dowding, Keith M. *Power*. Minneapolis, MN: University of Minnesota Press, 1996.

Drucker, Peter. *Post-Capitalist Society*. New York: Harper, 1993.
Dube, Arindrajit. "Minimum Wages and the Distribution of Family Incomes." Working paper, University of Massachusetts Amherst and IZA, 2013.
DuBois, Ellen Carol. *Women's Suffrage and Women's Rights*. New York: New York University Press, 1998.
Dunn, James J. "Letter to the President." In *People and the President: America's Conversations with FDR*, edited Lawrence W. Levine and Cornelia R. Levine, 69. Boston: Beacon Press, 2002.
Ellison, Nicole B., Charles Steinfield, and Cliff Lampe. "The Benefits of Facebook 'Friends': Social Capital and College Students' Use of Online Social Network Sites." *Journal of Computer-Mediated Communication* 12 (2007): 1143–68.
Elwell, Craig K. *Inflation and the Real Minimum Wage: A Fact Sheet*. Washington, DC: Congressional Research Service, 2014.
Ennis, James G., and Richard Schreuer. "Mobilizing Weak Support for Social Movements: The Role of Grievance, Efficacy, and Cost." *Social Forces* 66, no. 2 (1987): 390–409. https://doi.org/10.1093/sf/66.2.390.
Eskew, Glenn T. *But for Birmingham: The Local and National Movements in the Civil Rights Struggle*. Chapel Hill: University of North Carolina Press, 1997.
Evans, Robert, and Jonathan P. Thomas. "Reputation and Experimentation in Repeated Games with Two Long-Run Players." *Econometrica* 65, no. 5 (1997): 1153–73.
Evers, Medgar. "I Speak as a Native Mississippian." Televised address, May 20, 1963. WBLT, Jackson, MS, reprinted as "The Years of Change Are Upon Us," Crisis, 1973, 191–94.
Fantasia, Rick. *Cultures of Solidarity: Consciousness, Action, and Contemporary American Workers*. Berkeley: University of California Press, 1988.
Felsenthal, Carol. *The Sweetheart of the Moral Majority: The Biography of Phyllis Schlafly*. New York: Knopf, 1981.
Ferree, Myra Max. "The Political Context of Rationality: Rational Choice Theory and Resource Mobilization." In *Frontiers in Social Movement Theory*, edited by Aldon D. Morris and Carol McClung Mueller, 29-53. New Haven, CT: Yale University Press, 1992. https://doi.org/10.1111/j.1475-682x.1985.tb00850.x.
Ferree, Myra Marx, and Frederick D. Miller. "Mobilization and Meaning: Toward an Integration of Social Psychological and Resource Perspectives on Social Movements." *Sociological Inquiry* 55, no.1 (1985): 38–61.
Fireman, Bruce, and William A. Gamson. "Utilitarian Logic in the Resource Mobilization Perspective." In *The Dynamics of Social Movements: Resource Mobilization, Social Control, and Tactics*, edited by Mayer N. Zald and John D. McCarthy, 8-44. Cambridge, MA: Winthrop, 1979.
Fischer, David Hackett. *Paul Revere's Ride*. Oxford: Oxford University Press, 1994.
Foer, Franklin. *World without Mind: The Existential Threat of Big Tech*. New York: Penguin Press, 2017.
Frank, Nathaniel. *Awakening: How Gays and Lesbians Brought Marriage Equality to America*. Cambridge, MA: Belknap Press of Harvard University Press, 2017.
Freeman, Jo. "The Origins of the Women's Liberation Movement." *American Journal of Sociology* 78, no. 4 (1973): 792–811.
Friedman, Thomas L. *Thank You for Being Late: An Optimist's Guide to Thriving in the Age of Accelerations*. New York: Farrar, Straus, and Giroux, 2016.
Gambetta, David. "Can We Trust Trust?" In *Trust: Making and Breaking Cooperative Relations*, edited by David Gambetta, 213-38. Oxford: Basil Blackwell, 1988.

Gamson, William. "The Social Psychology of Collective Action." In *Frontiers in Social Movement Theory*, edited by Aldon D. Morris and Carol McClung Mueller, 53-76. New Haven, CT: Yale University Press, 1992.

Ganz, Marshall. *Why David Sometimes Wins: Leadership, Organization, and Strategy in the California Farm Worker Movement*. New York: Oxford University Press, 2009.

Garrow, David. *Bearing the Cross: Martin Luther King Jr. and the Southern Christian Leadership Conference*. New York: William Morrow, 2004.

Gelb, Joyce. "Social Movement 'Success': A Comparative Analysis of Feminism in the United States and the United Kingdom." In *The Women's Movements of the United States and Western Europe*, edited by Mary Fainsod Katzenstein and Carol McClurg Muller, 267-89. Philadelphia: Temple University Press, 1987.

Gelb, Joyce, and Marian Lief Palley. *Women and Public Policies: Reassessing Gender Politics*. Charlottesville: University of Virginia Press, 1996.

Gibbons, Robert. "Trust in Social Structures: Hobbes and Coase Meet Repeated Games." In *Trust in Society*, edited by Karen S. Cook, 332–49. New York: Russell Sage Foundation, 2001.

Gilder, George. "Metcalfe's Law and Legacy." *Forbes*, September 13, 1993.

Gittell, Ross, and Avis Vidal. *Community Organizing: Building Social Capital as a Development Strategy*. Thousand Oaks, CA: Sage, 1998.

Glaeser, Edward L., David I. Laibson, José A. Scheinkman, and Christine L. Soutter. "Measuring Trust." *Quarterly Journal of Economics* 115, no. 3 (2000): 811–46.

Goffman, Erving. *Frame Analysis*. Cambridge, MA: Harvard University Press, 1974.

Goldscheid, Julie. "Gender Neutrality, the Violence against Women Frame, and Transformative Reform." *University of Missouri at Kansas City Law Review* 82, no. 3 (2014): 623–60.

Gonzalez, Juan, and Joseph Torres. *News for All the People: The Epic Story of Race and the American Media*. London: Verso, 2011.

Granovetter, Mark S. "The Strength of Weak Ties." *American Journal of Sociology* 78, no. 6 (1973): 1360–80.

Grootaert, Christiaan, and Thierry van Bastelaer. "Understanding and Measuring Social Capital: An Integrated Questionnaire." World Bank working paper no. 18, Washington, DC: World Bank, 2004. Accessed April 3, 2019. https://openknowledge.worldbank.org/bitstream/handle/10986/15033/281100PAPER0Measuring0social0capital.pdf?sequence=1.

Halbert, Deborah. "Two Faces of Disintermediation: Corporate Control or Accidental Anarchy." *Michigan State Law Review* 83 (2006): 86–89.

Hall, Doug, and David Cooper. "How Raising the Federal Minimum Wage Would Help Working Families and Give the Economy a Boost." *Economic Policy Institute* 341 (2012): 1–19.

Hamilton, Charles V. *The Bench and the Ballot: Southern Federal Judges and Black Voters*. Oxford: Oxford University Press, 1973.

Hampton, Keith, and Barry Wellman. "Neighboring in Netville: How the Internet Supports Community and Social Capital in a Wired Suburb." *City & Community* 2, no. 4 (2003): 277–311.

Handler, Joel. *Social Movements and the Legal System: A Theory of Law Reform and Social Change*. New York: Academic Press, 1978.

Hanifan, L.J. "The Rural School Community Center." *Annals of the American Academy of Political and Social Sciences* 67 (1916): 130–38.

Hardin, Garrett. "The Tragedy of the Commons." *Science* 162, no. 3859 (1968): 1243–48.

Hardin, Russell. *Trust and Trustworthiness*. New York: Russell Sage Foundation, 2002.
Heath, Chip, and Dan Heath. *Switch: How to Change Things When Change Is Hard*. New York: Broadway Books, 2010.
Henkin, David M. *The Postal Age: The Emergence of Modern Communication in Nineteenth Century America*. Chicago: University of Chicago Press, 2006.
Hipp, John R., and Andrew J. Perrin. "The Simultaneous Effect of Social Distance and Physical Distance on the Formation of Neighborhood Ties." *City and Community* 8, no. 1 (2009): 5–25.
Hoffman, Elizabeth, Kevin McCabe, and Vernon L. Smith. "Social Distance and Other-Regarding Behavior in Dictator Games." *American Economic Review* 86, no. 3 (1996): 653–60.
Howe, Daniel Walker. *What God Hath Wrought: The Transformation of America, 1815–1848*. Oxford: Oxford University Press, 2007.
Howe, Jeff. *Crowdsourcing: Why the Power of the Crowd Is Driving the Future of Business*. New York: Crown Business, 2008.
Huckshorn, Robert J. "National Committee Leadership of State and Local Parties." In *Politics, Professionalism and Power: Modern Party Organization and the Legacy of Ray C. Bliss*, edited by John C. Green, 34–47. Lanham, MD: University Press of America, 1994.
Hutchinson, Darren Lenard. "Preventing Balkanization or Facilitating Racial Domination: A Critique of the New Equal Protection." *Virginia Journal of Social Policy & Law* 22, no. 1 (2015): 3–69.
Hyland, William. *George Gershwin: A New Biography*. Westport, CT: Praeger Publishers, 2003.
Information Commissioner's Office, *Investigation into the Use of Data Analytics in Political Campaigns: A Report to Parliament*. November 6, 2018.
Issenberg, Sasha. *The Victory Lab: The Secret Science of Winning Campaigns*. New York: Crown, 2012.
Jackson, Maurice, Eleanora Petersen, James Bull, Sverre Monsen, and Patricia Richmond. "The Failure of an Incipient Social Movement." *Pacific Sociological Review* 3, no. 1(1960): 35–40.
Jameson, J. Franklin. *The American Revolution Considered as a Social Movement*. Princeton, NJ: Princeton University Press, 1926.
Jenkins, Craig J., and William Form. "Social Movements and Social Change." In *The Handbook of Political Sociology: States, Civil Societies and Globalization*, edited by Thomas Janoski, 331-49. New York: Cambridge University Press, 2005.
Jensen, Merrill. *The Founding of a Nation: A History of the American Revolution, 1763–1776*. Cambridge, MA: Hackett, 1968.
John, Richard R. *Spreading the News: The American Postal System from Franklin to Morse*. Cambridge, MA: Harvard University Press, 1998.
"Johnson Conversation with Martin Luther King on January 15, 1965." Miller Center of Public Affairs, University of Virginia. Transcript and WAV MP3 audio, 13:43. Accessed June 17, 2019. https://millercenter.org/the-presidency/secret-white-house-tapes/conversation-martin-luther-king-and-office-secretary.
Johnson, Lyndon B. "Special Message to the Congress: The American Promise." Address, U.S. Capitol, Washington, DC. March 15, 1965.
Johnston, Hank, Enrique Laraña, and Joseph R. Gusfield. "Identities, Grievances and New Social Movements." In *New Social Movements: From Ideology to Identity*, edited by Enrique Laraña, Henry Johnston, and Joseph R. Gusfield, 3-35. Philadelphia: Temple University Press, 1994.

Jones, Jason J., Robert M. Bond, Eytan Bakshy, Dean Eckles, and James H. Fowler. "Social Influence and Political Mobilization: Further Evidence from a Randomized Experiment in the 2012 U.S. Presidential Election." *PLoS ONE* 12, no. 4 (2017): e0173851.

Judis, John B. *The Paradox of American Democracy: Elites, Special Interests, and the Betrayal of Public Trust*. New York: Routledge, 2000.

Kadushin, Charles. *Understanding Social Networks: Theories, Concepts, and Findings*. Oxford: Oxford University Press, 2012.

Kahan, Dan M. "The Logic of Reciprocity: Trust, Collective Action, and Law." *Michigan Law Review* 102, no. 1 (2003): 71–103.

Kahneman, Daniel, and Amos Tversky. "Choices, Values and Frames." *American Psychologist* 39, no. 4 (1984): 341–50.

Kasparov, Garry. *Deep Thinking: Where Machine Intelligence Ends and Human Creativity Begins*. New York: Public Affairs, 2017.

Kavanaugh, Andrea, Debbie Denise Reese, John M. Carroll, and Mary Beth Rosson. "Weak Ties in Networked Communities." In *Communities and Technologies*, edited by M. Huysman, E. Wenger, and V. Wulf, 265–86. Dordrecht: Springer, 2003.

Kazin, Michael, Rebecca Edwards, and Adam Rothman. *Princeton Encyclopedia of American Political History*. Princeton, NJ: Princeton University Press, 2009.

Kelly, Kevin. *The Inevitable: Understanding the Technological Forces That Will Shape Our Future*. New York: Penguin, 2016.

Kittilson, Miki Caul, and Russell J. Dalton. "Virtual Civil Society: The New Frontier of Social Capital?" *Political Behavior* 33, no. 4 (2011): 625–44.

Klandermans, Bert, and Dirk Oegema. "Potentials, Networks, Motivations, and Barriers: Steps towards Participation in Social Movements." *American Sociological Review* 52, no. 4 (1987): 519–31.

Klinkner, Phillip A. "A Comparison of Out-Party Leaders: Ray Bliss and Bill Brock." In *Politics, Professionalism and Power: Modern Party Organization and the Legacy of Ray C. Bliss*, edited by John C. Green, 135-48. Lanham, MD: University Press of America, 1994.

Knack, Stephen. "Social Capital and the Quality of Government: Evidence from the States." *American Journal of Political Science* 46, no. 4 (2002): 772–85.

Knack, Stephen, and Phillip Keefer. "Does Social Capital Have an Economic Payoff? A Cross-Country Investigation." *Quarterly Journal of Economics* 112, no. 4 (1997): 1251–88.

Kreps, David M. *Game Theory and Economic Modeling*. New York: Clarendon Press, 1990.

Lanier, Jaron. *Ten Arguments for Deleting Your Social Media Accounts Right Now*. New York: Henry Holt, 2018.

LaPorta, Rafael, Florencio Lopez-de-Silane, Andrei Shleifer, and Robert W. Vishny. "Trust in Large Organizations." *American Economic Review Papers & Proceedings* 87, no. 2 (1997): 333–38.

Lazer, David M.J., Matthew A. Baum, Yochai Benkler, Adam J. Berinsky, Kelly M. Greenhill, Filippo Menczer, Miriam J. Metzger, Brendan Nyhan, Gordon Pennycook, David Rothschild, Michael Schudson, Steven A. Sloman, Cass R. Sunstein, Emily A. Thorson, Duncan J. Watts, and Jonathan L. Zittrain. "The Science of Fake News." *Science* 359 (2018): 1094–96.

Lenthall, Bruce. *Radio's America: The Great Depression and the Rise of Modern Mass Culture*. Chicago: University of Chicago Press, 2007.

Lepore, Jill. *These Truths: A History of the United States*. New York: W.W. Norton, 2018.

Liebman, James S., and Charles F. Sabel. "A Public Laboratory Dewey Barely Imagined: The Emerging Model of School Governance and Legal Reform." *New York University Review of Law and Social Change* 28 (2003): 184–304.

Lilla, Mark. *The Once and Future Liberal: After Identity Politics*. New York: Harper, 2017.

Lo, Clarence Y.H. *Small Property versus Big Government: Social Origins of the Property Tax Revolt*. Berkeley: University of California Press, 1990.

Loomis, James L. "Communication, the Development of Trust, and Cooperative Behavior." *Human Relations* 12 (1959): 305–15. http://dx.doi.org/10.1177/001872675901200402.

Los Angeles Alliance for a New Economy. "A Tale of Two Cities: How Long Beach's Investment in Downtown Tourism Has Contributed to Poverty Next Door." September 10, 2009. Accessed April 4, 2019. http://laane.org/downloads/B335P3112C.pdf.

Los Angeles Alliance for a New Economy. "Raise LA: How Good Hotel Jobs Will Boost Local Businesses, Strengthen Neighborhoods, and Renew Our Economy." July 2013. Accessed April 4, 2019. https://www.laane.org/wp-content/uploads/2013/08/Raise-LA-Report-FINAL-SMALL.pdf.

Lowrey, Annie. *Give People Money: How a Universal Basic Income Would End Poverty, Revolutionize Work, and Remake the World*. New York: Crown, 2018.

Lundby, Knut. "Introduction: Digital Storytelling, Mediated Stories." In *Digital Storytelling, Mediated Stories: Self-Representation in New Media*, edited by Knut Lundby, 1-15. New York: Peter Lang Publishing, 2008.

Magdol, Edward. *The Antislavery Rank and File: A Social Profile of the Abolitionists' Constituency*. New York: Greenwood Press, 1986.

Manis, Andrew M. *A Fire You Can't Put Out: The Civil Rights Life of Birmingham's Reverend Fred Shuttlesworth*. Tuscaloosa: University of Alabama Press, 2001.

Mannheim, Karl. *Man and Society: In an Age of Reconstruction*. Translated by Edward Shils. New York: Harcourt, Brace & World, 1940.

Mansbridge, Jane. *Why We Lost the ERA*. 2nd ed. Chicago: University of Chicago Press, 1986.

Margetts, Helen, and Peter John. *Political Turbulence: How Social Media Shape Collective Action*. Princeton, NJ: Princeton University Press, 2016.

"Mark Zuckerberg, in His Own Words: The CNN Interview." CNN. March 21, 2018. Accessed April 4, 2019. http://money.cnn.com/2018/03/21/technology/mark-zuckerberg-cnn-interview-transcript/index.html.

Mazar, Nina, On Amir, and Dan Ariely. "The Dishonesty of Honest People: A Theory of Self-Concept Maintenance." *Journal of Marketing Research* 45 (December 2008): 633–44.

Mazzone, Jason. "Toward a Social Capital Theory of Law: Lessons from Collaborative Reproduction." *Santa Clara Law Review* 39, no. 1 (1998): 1–77.

McAdam, Doug. "Culture and Social Movements." In *New Social Movements: From Ideology to Identity*, edited by Enrique Laraña, Hank Johnston, and Joseph R. Gusfield, 36-57. Philadelphia: Temple University Press, 1994.

McAdam, Doug. "Recruitment to High-Risk Activism: The Case of Freedom Summer." *American Journal of Sociology* 92, no. 1 (1986): 64–90.

McAdam, Doug, John D. McCarthy, and Mayer N. Zald. "Social Movements." In *Handbook of Sociology*, edited by Neil J. Smelser, 36-57. Newbury Park: Sage, 1988.

McAfee, Andrew, and Erik Brynjolfsson. *Machine, Platform, Crowd: Harnessing Our Digital Future*. New York: W.W. Norton, 2017.

McAlevey, Jane. *No Shortcuts: Organizing for Power in the New Gilded Age*. Oxford: Oxford University Press, 2016.

McCarthy, John D. "Pro-Life and Pro-Choice Mobilization: Infrastructure Deficits and New Technologies." In *Social Movements in an Organizational Society*, edited by Mayer N. Zald and John D. McCarthy, 695-738. New Brunswick, NJ: Transaction Books, 1987.

McCarthy, John D., and Mark Wolfson. "Consensus Movements, Conflict Movements, and the Cooptation of Civic and State Infrastructures." In *Frontiers in Social Movement Theory*, edited by Aldon D. Morris and Carol McClurg Mueller, 273-95. New Haven, CT: Yale University Press, 1992.

McCarthy, John D., and Mark Wolfson. "Resource Mobilization by Local Social Movement Organizations: Agency, Strategy, and Organization in the Movement against Drunk Driving." *American Sociological Review* 61, no. 6 (1996): 1070–88.

McCarthy, John D., and Mayer N. Zald. "Resource Mobilization and Social Movements: A Partial Theory." *American Journal of Sociology* 82, no. 6 (1977): 1212–41.

McCarthy, John D., and Mayer N. Zald. "The Trend of Social Movements in America: Professionalization and Resource Mobilization." Working paper, Center for Research on Social Organization, Ann Arbor, Michigan, 1973.

McLuhan, Marshall. *Understanding Media: The Extensions of Man*. Edited by W. Terrence Gordon. Berkeley: Gingko Press, 1994.

McNamee, Roger. *Zucked: Waking Up to the Facebook Catastrophe*. New York: Penguin Press, 2019.

McPherson, Miller, Lynn Smith-Lovin, and James Cook. "Birds of a Feather: Homophily in Social Networks." *Annual Review of Sociology* 27 (August 2001): 415–44. https://doi.org/10.1146/annurev.soc.27.1.415.

Melucci, Alberto. "A Strange Kind of Newness: What's 'New' in New Social Movements." In *New Social Movements: From Ideology to Identity*, edited by Enrique Laraña, Henry Johnston, and Joseph R. Gusfield, 101-29. Philadelphia: Temple University Press, 1994.

Merrill, Walter M., and Louis Ruchames, eds. *Letters of William Lloyd Garrison*. Cambridge, MA: Harvard University Press, 1971.

Mettler, Suzanne. *Soldiers to Citizens: The G.I. Bill and the Making of the Greatest Generation*. Oxford: Oxford University Press, 2005.

Mill, John Stuart. *On Liberty*. 2nd ed. Boston: Ticknor and Fields, 1863.

Miller, John C., *Origins of the American Revolution*. Redwood City, CA: Stanford University Press, 1959.

Minimum Wage Study Commission. *Report of the Minimum Wage Study Commission*. Ann Arbor: Minimum Wage Study Commission, 1981.

Minkoff, Debra C. *Organizing for Equality: The Evolution of Women's and Racial-Ethnic Organizations in America, 1955–1985*. New Brunswick, NJ: Rutgers University Press, 1995.

Morgan, Edmund S., ed. *Prologue to Revolution: Sources and Documents on the Stamp Act Crisis, 1764–1766*. Chapel Hill: University of North Carolina Press, 1959.

Morgan, Edmund S., and Helen Morgan. *The Stamp Act Crisis: Prologue to Revolution*. Chapel Hill: University of North Carolina Press, 1995.

Morozov, Evgeny. *The Net Delusion: The Dark Side of Internet Freedom*. New York: Public Affairs, 2012.

Morris, Aldon. *The Origins of the Civil Rights Movement: Black Communities Organizing for Change*. New York: Free Press, 1984.

Mullen, Mary K. "The Violence Against Women Act: A Double-Edged Sword for Native Americans, Their Rights, and Their Hopes of Regaining Cultural Independence." *St. Louis University Law Journal* 61 (2017): 811–34.

National Coalition of Anti-Violence Programs. *Lesbian, Gay, Bisexual, Transgender, Queer and HIV-Affected Intimate Partner Violence, 2011*. New York, 2012. Accessed April 1, 2019, https://avp.org/wp-content/uploads/2017/04/2011_NCAVP_IPV_Report.pdf.

National Task Force to End Sexual and Domestic Violence Against Women. "Summary of Changes from VAWA Reauthorization 2013." 2013. Accessed April 1, 2019. http://www.ncdsv.org/images/NTFESDVAW_SummaryOfChangesFromVAWAreauthorization2013.pdf.

Newman, Richard S. *The Transformation of American Abolitionism: Fighting Slavery in the Early Republic*. Chapel Hill: University of North Carolina Press, 2002.

Noah, Timothy. *The Great Divergence: America's Growing Inequality Crisis and What We Can Do about It*. New York: Bloomsbury Press, 2012.

Norris, Pippa. *Democratic Phoenix: Reinventing Political Activism*. Cambridge: Cambridge University Press, 2002.

Obar, Jonathan A., Paul Zube, and Clifford Lampe. "Advocacy 2.0: An Analysis of How Advocacy Groups in the United States Perceive and Use Social Media as Tools for Facilitating Civic Engagement and Collective Action." *Journal of Information Policy* 2 (2012): 1–25.

Oberschall, Anthony. *Social Conflict and Social Movements*. Englewood Cliffs, NJ: Prentice-Hall, 1973.

Olivares, Mariela. "Battered by Law: The Political Subordination of Immigrant Women." *American University Law Review* 64, no. 2 (2014): 231–83.

Oliver, Pamela, and Mark Furman. "Contradictions between National and Local Organizational Strength: The Case of the John Birch Society." In *International Social Movement Research*, 155–77. Greenwich, CT: JAI Press, 1989.

Oliver, Pamela E., and Gerald Marwell. "Mobilizing Technologies for Collective Action." In *Frontiers in Social Movement Theory*, edited by Aldon D. Morris and Carol McClurg Mueller, 251-72. New Haven, CT: Yale University Press, 1992.

Olson, Keith W. *The G.I. Bill, the Veterans and the Colleges*. Lexington: University Press of Kentucky, 1974.

Olson, Mancur. *The Logic of Collective Action: Public Goods and the Theory of Groups*. Cambridge, MA: Harvard University Press, 1965.

O'Neil, Cathy. *Weapons of Math Destruction: How Big Data Increases Inequality and Threatens Democracy*. New York: Broadway Books, 2016.

Orbell, John M., Alphons J. Van de Kragt, and Robyn M. Dawes. "Explaining Discussion-Induced Cooperation." *Journal of Personality and Social Psychology* 54, no. 5 (1988): 811–19.

O'Shaughnessy, Nicholas. *Marketing the Third Reich: Persuasion, Packaging and Propaganda*. London: Routledge, 2018.

Ostrom, Elinor. *Governing the Commons: The Evolution of Institutions for Collective Action*. Cambridge: Cambridge University Press, 1990.

Ostrom, Elinor, and T.K. Ahn. "The Meaning of Social Capital and Its Link to Collective Action." In *Handbook of Social Capital*, edited by Gert T. Svendsen and Gunnar L. Svendsen 17-35. Cheltenham: Edward Elgar, 2009.

Painter, Nell Irvin. *Standing at Armageddon: A Grassroots History of the Progressive Era*. New York: Norton, 2008.

Palfrey, John, and Urs Gasser. *Born Digital: Understanding the First Generation of Digital Natives*. New York: Basic Books, 2008.

Perry, Martha W., and Alexandra V. Ricca. "Instant Messaging: Virtual Propinquity for Health." *Promotion and Education* 13, no. 3 (2006): 211–12.

Piketty, Thomas. *Capital in the 21st Century*. Translated by Arthur Goldhammer. Cambridge, MA: Belknap Press of Harvard University Press, 2014.

Piketty, Thomas, Emmanuel Saez, and Gabriel Zucman. "Distributional National Accounts: Methods and Estimates for the United States." Working paper 22945, National Bureau of Economic Research, 2016.

Pitkin, R.B. "How the First GI Bill Was Written (Part II)." *American Legion Magazine*, February 1969.

Polsby, Nelson W. *Presidential Elections: Strategies and Structures of American Politics.* Lanham, MD: Rowman & Littlefield, 2008.

Porta, Donatella Della, and Mario Diani. *Social Movements: An Introduction*. Malden, MA: Blackwell, 2006.

Powell, Lewis F., Jr. "Attack on American Free Enterprise System." Lewis F. Powell Jr. to Eugene B. Syndor, Jr., memorandum, August 23, 1971. Accessed June 14, 2019. https://lawdigitalcommons.bc.edu/cgi/viewcontent.cgi?article=1078&context=darter_materials.

Purdy, Jedediah. *For Common Things: Trust and Commitment in America Today*. New York: Vintage Books, 2010.

Putnam, Robert D. "Bowling Alone: America's Declining Social Capital." *Journal of Democracy* 6 (January 1995): 65–78.

Putnam, Robert D. *Bowling Alone: The Collapse and Revival of American Community*. New York: Simon & Schuster, 2000.

Putnam, Robert D. "*E Pluribus Unum*: Diversity and Community in the 21st Century: The 2006 Johan Skytte Prize Lecture." *Scandinavian Political Studies* 30, no.2 (2007): 137–74.

Putnam, Robert D., Robert Leonardi, and Raffaella Nanetti. *Making Democracy Work: Civic Traditions in Modern Italy*. Princeton, NJ: Princeton University Press, 1993.

Quan-Haase, Anabel, and Barry Wellman. "How Does the Internet Affect Social Capital?" In *Social Capital and Information Technology*, edited by Marleen Huysman and Volker Wulf, 113-32. Cambridge, MA: MIT Press, 2004.

Quarles, Benjamin. *Black Abolitionists*. New York: Oxford University Press, 1969.

Raines, Howell. *My Soul Is Rested: Movement Days in the Deep South Remembered.* New York: Penguin Books, 1983.

Ramsay, David. *The History of the American Revolution*. Vol. 2. New York: James J. Wilson, 1811.

Rawls, John. *A Theory of Justice.* Cambridge, MA: Harvard University Press, 1971.

Risley, Ford. *Abolition and the Press: The Moral Struggle against Slavery*. Chicago: Northwestern University Press, 2008.

Roe, Merwin. *Speeches and Letters of Abraham Lincoln, 1832–1865*. London: J.M. Dent & Sons, 1907.

Roosevelt, Franklin D. FDR Letter to Ralph W. Farrell. Democratic National Committee Campaign Correspondence, 1928–1933. Franklin D. Roosevelt Library, Hyde Park, New York.

Roosevelt, Franklin D. First inaugural address. March 4, 1933. The American Presidency Project. Edited by John Woolley and Gerhard Peters. https://www.presidency.ucsb.edu/documents/inaugural-address-8.

Roosevelt, Franklin D. "On the Banking Crisis." March 12, 1933. Transcript and Adobe Flash MP3 audio. American Rhetoric. http://www.americanrhetoric.com/speeches/fdrfirstfiresidechat.html.
Rose, Carol M. "Trust in the Mirror of Betrayal." *Boston University Law Review* 75 (1995): 531–58.
Sabatini, Fabio, and Francesco Sarracino. "E-participation, Social Capital and the Internet." FEEM working paper 081.2014, 2014.
Sabel, Charles F. "Constitutional Ordering in Historical Context." In *Games in Hierarchies and Networks: Analytical and Empirical Approaches to the Study of Governance Institutions*, edited by Fritz W. Scharpf, 65–123. Boulder, CO: Westview Press, 1993.
Sajuria, Javier, Jennifer van Heerde-Hudson, David Hudson, Niheer Dasandi, and Yannis Theocharis. "Tweeting Alone? An Analysis of Bridging and Bonding Social Capital in Online Networks." *American Politics Research* 43 (2015): 708–38.
Sally, David. "Conversation and Cooperation in Social Dilemmas: A Meta-Analysis of Experiments from 1958 to 1992." *Rationality and Society* 7, no. 1 (1995): 58–92.
Sally, David. "Game Theory Behaves." *Marquette Law Review* 87 (2004): 783–93.
Sandoz, Ellis, and Cecil V. Crabb, Jr. *A Tide of Discontent: The 1980 Elections and Their Meaning*. Washington, DC: Congressional Quarterly Press, 1981.
Schmitt, John. *Why Does the Minimum Wage Have No Discernible Effect on Employment?* Washington, DC: Center for Economic and Policy Research, 2013.
Schwarz, Michael, and Shuva Paul. "Resource Mobilization versus the Mobilization of People: Why Consensus Movements Cannot Be Instruments of Change." In *Frontiers in Social Movement Theory*, edited by Aldon D. Morris and Carol McClurg Mueller, 205–23. New Haven, CT: Yale University Press, 1992.
Servon, Lisa J. "Credit and Social Capital: The Community Development Potential of U.S. Microenterprise Programs." *Housing Policy Debate* 9, no. 1 (1998): 115–49. doi:10.1080/10511482.1998.9521288.
Shalot, Tali. *The Optimism Bias: A Tour of the Irrationally Positive Brain*. New York: Random House, 2011.
Shapiro, Carl, and Hal R. Varian. *Information Rules: A Strategic Guide to the Network Economy*. Boston: Harvard Business School Press, 1999.
Shirky, Clay. *Cognitive Surplus: Creativity and Generosity in a Connected Age*. New York: Penguin Press, 2010.
Shirky, Clay. *Here Comes Everybody: The Power of Organizing without Organizations*. New York: Penguin Press, 2008.
Sitaraman, Ganesh. *The Crisis of the Middle-Class Constitution: Why Inequality Threatens Our Republic*. New York: Knopf, 2017.
Sklar, Kathryn Kish. "Women Who Speak for an Entire Nation: American and British Women Compared at the World Anti-Slavery Convention, London 1840." *Pacific Historical Review* 59, no. 4 (1990): 453–99.
Skocpol, Theda. *Diminished Democracy: From Membership to Management in American Civil Life*. Norman: University of Oklahoma Press, 2003.
Skocpol, Theda. "Targeting within Universalism: Politically Viable Policies to Combat Poverty in the United States." In *The Urban Underclass*, edited by Christopher Jencks and Paul E. Peterson, 411–36. Washington, DC: Brookings Institution Press, 1991.
Skocpol, Theda. *States and Social Revolutions: A Comparative Analysis of France, Russia, and China*. 2nd ed. Cambridge, UK: Cambridge University Press, 2015.

Slaughter, Anne-Marie. *The Chessboard and the Web: Strategies of Connection in a Networked World.* New Haven, CT: Yale University Press, 2017.

Snow, David A., and Robert D. Benford, "Master Frames and Cycles of Protest." In *Frontiers in Social Movement Theory*, edited by Aldon D. Morris and Carol McClurg Mueller. New Haven, CT: Yale University Press, 1992.

Snow, David A., E. Burke Rochford, Jr., Steven K. Worden, and Robert D. Benford, "Frame Alignment Processes, Micromobilization, and Movement Participation." *American Sociological Review* 51 (1986): 464–81.

Snow, David A., Louis A. Zurcher, Jr., and Sheldon Ekland-Olson. "Social Networks and Social Movements: A Microstructural Approach to Differential Recruitment." *American Sociological Review* 45, no. 5 (1980): 787–801.

Staggenborg, Suzanne. "The Consequences of Professionalization and Formalization in the Pro-Choice Movement." *American Sociological Review* 53, no. 4 (1988): 585–605.

Standage, Tom. *Writing on the Wall: Social Media–The First 2000 Years.* New York: Bloomsbury, 2013.

Standage, Tom. *The Victorian Internet: The Remarkable Story of the Telegraph and the Nineteenth Century's On-line Pioneers.* New York: Bloomsbury, 1999.

Stanton, Elizabeth Cady. *Eighty Years or More: Reminiscences 1815 to 1897.* Whitefish, MT: Kessinger, 2004.

Starr, Paul. *The Creation of the Media: Political Origins of Modern Communications.* New York: Basic Books, 2005.

Stolle, Dietland, and Marc Hooghe. "Review Article: Inaccurate, Exceptional, One-Sided or Irrelevant? The Debate about the Alleged Decline of Social Capital and Civic Engagement in Western Societies." *British Journal of Political Science* 35 (2004): 149–67.

Story, Joseph. *Commentaries on the Constitution.* Boston: Hilliard Gray, 1833.

Sullivan, Andrew. *Virtually Normal: An Argument about Homosexuality.* New York: Vintage Books, 1995.

Sunstein, Cass R. *How Change Happens.* Cambridge, MA: MIT Press, 2019.

Sunstein, Cass R. *#Republic: Divided Democracy in the Age of Social Media.* Princeton, NJ: Princeton University Press, 2017.

Surowiecki, James. *The Wisdom of Crowds.* New York: Random House, 2005.

Taleb, Nassim Nicholas. *Skin in the Game: Hidden Asymmetries in Daily Life.* New York: Random House, 2018.

Tarrow, Sidney G. *The Power in Movement: Social Movements, Collective Action, and Politics* New York: Cambridge University Press, 1994.

Taylor, Astra. *The People's Platform: Taking Back Power and Culture in the Digital Age.* New York: Metropolitan Books, 2014.

Taylor, Verta, and Nancy Whittier. "Collective Identity in Social Movement Communities: Lesbian Feminist Mobilization." In *Frontiers in Social Movement Theory*, edited by Aldon D. Morris and Carol McClung Mueller, 104–29. New Haven, CT: Yale University Press, 1992.

Tebbel, John. *The Media in America.* New York: Crowell, 1974.

Tilly, Charles. "Repertoires of Contention in America and Britain, 1750–1830." In *The Dynamics of Social Movements: Resource Mobilization, Social Control, and Tactics*, edited by Mayer N. Zald and John D. McCarthy, 126-55. Cambridge, MA: Winthrop, 1979.

Tilly, Charles. "The Web of Contention in Eighteenth-Century Cities." In *Class Conflict and Collective Action*, edited by Louise A. Tilly and Charles Tilly, 27–51. Beverly Hills, CA: Sage, 1981.

Tocqueville, Alexis de. *Democracy in America*. Translated by Harvey C. Mansfield and Delba Winthrop. Chicago: University of Chicago Press, 2000.

Topolsky, Mary. "Common Cause? The Illusion of Participating." *Worldview* 17, no. 3 (1974): 35–39.

Torres, Sasha. *Black, White, and in Color: Television and Black Civil Rights*. Princeton, NJ: Princeton University Press, 2003.

Tufekci, Zeynep. *Twitter and Tear Gas: The Power and Fragility of Networked Protests*. New Haven, CT: Yale University Press, 2016.

Tversky, Amos, and Daniel Kahneman. "Availability: A Heuristic for Judging Frequency and Probability." *Cognitive Psychology* 5, no. 2 (1973): 207–32.

Tyler, Tom R. *Why People Obey the Law*. Princeton, NJ: Princeton University Press, 2006.

U.S. Department of Labor, Bureau of Labor Statistics. "Union Members Summary." Last modified January 23, 2015. Accessed April 3, 2019. http://www.bls.gov/news.release/union2.nr0.htm.

Uslaner, Eric M. *The Moral Foundations of Trust*. Cambridge: Cambridge University Press, 2002.

Uslaner, Eric M. "Segregation, Mistrust and Minorities." *Ethnicities* 10, no. 4 (2010): 415–34.

Verbrugge, Lois M. "The Structure of Adult Friendship Choices." *Social Forces* 56, no. 2 (1977): 576–97.

Viguerie, Richard A., and David Franke. *America's Right Turn: How Conservatives Used New and Alternative Media to Take Power*. Chicago: Bonus Books, 2004.

Villalón, Roberta. "Violence against Immigrants in a Context of Crisis: A Critical Migration Feminist of Color Analysis." *Journal of Social Distress and the Homeless* 24, no. 3 (May 2015): 116–39. doi:10.1179/105307789 15z.000000000017.

Villalón, Roberta. *Violence against Latina Immigrants: Citizenship, Inequality, and Community*. New York: NYU Press, 2010.

Waddock, Sandra, Steven Waddell, and Paul S. Gray. "The Transformational Change Challenge of Memes: The Case of Marriage Equality in the United States." *Business & Society* (December 10, 2018): 1–31.

Weinstein, Neil D. "Unrealistic Optimism about Future Life Events." *Journal of Personality and Social Psychology* 39, no. 5 (1980): 806–20.

Weir, Margaret, and Marshall Ganz. "Reconnecting People and Politics." In *The New Majority: Toward a Popular Progressive Politics*, edited by Stanley B. Greenburg and Theda Skocpol, 149–71. New Haven, CT: Yale University Press, 1997.

Wellman, Barry. "Computer Networks as Social Networks." *Science* 293 (September 2001): 2031–34.

Wellman, Judith. *The Road to Seneca Falls: Elizabeth Cady Stanton and the First Woman's Rights Convention*. Urbana: University of Illinois Press, 2004.

West, Geoffrey. *Scale: The Universal Laws of Growth, Innovation, Sustainability, and the Pace of Life in Organisms, Cities, Economies, and Companies*. New York: Penguin Press, 2017.

Westen, Drew. *The Political Brain: The Role of Emotion in Deciding the Fate of the Nation*. New York: Public Affairs, 2007.

Whitman, Gordon. *Stand Up! How to Get Involved, Speak Out, and Win in a World on Fire*. New York: Penguin Random House, 2018.

Winter, Alice Ames. *The Business of Being a Club Woman*. New York: Century, 1925.

Wolfson, Evan. "Samesex Marriage and Morality: The Human Rights Vision of the Constitution." Thesis, Harvard University Law School, 1983. Accessed April 9, 2019. http://freemarry.3cdn.net/73aab4141a80237ddf_kxm62r3er.pdf.

Wood, Gordon S. *The Creation of the American Republic* 1776–1787. Chapel Hill: University of North Carolina Press, 1998.

World Bank. *World Development Report 2000/2001: Attacking Poverty*. Oxford: Oxford University Press, 2001.

Wroth, Lawrence C., *The Colonial Printer.* Portland, ME: Southworth-Anthoesen Press, 1938.

Wyrwoll, Claudia. "Social Media: Fundamentals, Models, and Ranking of User-Generated Content." PhD diss., Hamburg University, 2014.

Zak, Paul J., and Stephen Knack. "Trust and Growth." *Economic Journal* 111 (April 2001): 295–321.

Zuboff, Shoshana. *The Age of Surveillance Capitalism: The Fight for a Human Future at the New Frontier of Power*. New York: Public Affairs, 2019.

Index